Oceano Atlantico

'America' c. 1600 Reproduced by permission
Library of Congress Geography and Maps Division

The Man who foiled a Jamestown Massacre

David Edmund Pace

Acknowledgements

The Author wishes to acknowledge the help afforded him by the Adam Mathews group in allowing him free access to their data base. He would also like to pay tribute to Dr David Ransome who transposed previously unexamined Farrar Papers without which this book could not have been written. Extracts and transcripts of documents which throw light on the Virginia Company quoted here are held at Magdalene College, Cambridge and are reproduced by permission of the Master and Fellows of the College with whom copyright remains. I would like to acknowledge the debt to Susan Myra Kingsbury owed by all who study the early history of Virginia. The author would also like to thank Professor Vanesa Harding, Dept of History, Classics and Archaeology at Birkbeck College, University of London for permission to access the project 'People and Place'. Jordan Landes, Research Librarian at University of London provided a very helpful list of references. The author would like to acknowledge the invaluable data on transatlantic voyages 1607-37 painstakingly compiled by Anne Stevens and the members of the Pace Society who over many years have established an extensive data base on the Pace family genealogy. In Chapter One I quote extensively from 'The English and Their History' by Professor Robert Tombs, (Allen Lane 2014), which was extremely helpful to understanding the religious context of the period. The image of three vessels arriving in the James River and the extremely helpful detailed map of the plantations was sourced from the e-book of 'The First Seventeen Years of Virginia 1607 – 1624' by Charles E. Hatch published as The Project Gutenberg. The NOAA National Ocean Service kindly provided the tide table for March 1622. I would like to thank Nicholas Riddle for giving me insight into the composition of the congregation at St James, Clerkenwell. I would also like to thank Anne Countess De La Warr and Christopher Whittick, Senior Archivist East Sussex County Council for advice on the De La War family archives. Finally I would finally like to pay tribute to Betty Chamberlayne, an eminent local historian, formerly of The Court House, Maisemore, Gloucestershire, who gave me much help in researching the history of the Pace family in the West of England.
Every effort has been made to contact all copyright holders of material included in this book. If any have been inadvertently overlooked, the author will be pleased to make the necessary arrangements at the first opportunity.

<u>Cover Illustration</u>
The Indian massacre of 1622 depicted as a woodcut by Matthaeus Merian 1628

Published by David Edmund Pace
Publishing partner: Paragon Publishing, Rothersthorpe
First published 2016
© David Edmund Pace 2016

The rights of David Edmund Pace to be identified as the author of this work have been asserted by him in accordance with the Copyright, Designs and Patents Act of 1988.

All rights reserved; no part of this publication may be reproduced, stored in a retrieval system, or transmitted in any form or by any means, electronic, mechanical, photocopying, recording or otherwise without the prior written consent of the publisher or a licence permitting copying in the UK issued by the Copyright Licensing Agency Ltd. www.cla.co.uk

ISBN 978-1-78222-481-5

Book design, layout and production management by Into Print
www.intoprint.net
+44 (0)1604 832149

Printed and bound in UK and USA by Lightning Source

THE MAN WHO FOILED A JAMESTOWN MASSACRE

THE LIFE AND TIMES OF RICHARD PACE OF PACE'S PAINES

DAVID EDMUND PACE

To Richard and Isabel Pace
Ancient Planters of Virginia

'QUO FATA FERUNT'

('Whither the Fates Carry Us')
(Bermuda Coat of Arms)

Wolves in Sheep's Clothing

Humans have a Janus-like capacity to show remarkable altruism and nobility and at other times to be base, selfish, and duplicitous. This darker refrain of character encompasses the art of the unethical person who persuades, radicalizes, grooms and dupes for some personal advantage. It is the trade of the confidence trickster. A common factor is that it involves deliberate deception to gain a trust which is eventually to be betrayed. The ultimate betrayal is when the game is played for life itself, such as the man who smiles at his brother to disarm before he smites, or who gives a Judas kiss.

Man is a social animal who interacts and transacts with others. When he interacts he needs to signal his intentions. Some signals are chosen and some are involuntary. Signals made in a casual meeting or during shared pastimes may be of less significance than when engaged in a more purposeful meeting which might involve a degree of negotiation. Here a gleam in the eye, the alacrity of response to a proposal or the urgency of voice may be vital clues to the keenness to agree a deal.

On the morning of the 22nd March 1621/22 such transactions were taking place in the townships and isolated settlements on both sides of the James River in Virginia. Relationships would have been outwardly friendly for each meeting was intended to secure a mutually beneficial exchange. The tribal visitors must have concealed any hint of their hidden agenda in order to lull their hapless victims into a false sense of security. At a chosen moment the mask of friendship slipped, faces changed from an agreeable expression to vengefulness, the soft language of discussion to aggressive war hoops and the body language to attack mode. The local Indians fell upon their unsuspecting hospitable hosts, seizing whatever was to hand to hack and murder the vulnerable and undefended settlers. Mercy was granted to none. The shrieks, cries and moans from the dying would have been followed by the dreadful silence of death and the gruesome desecration of the corpses. There is a particular term to describe these predators when they work as a pack. They are known as 'Wolves in Sheep's Clothing'

In total nearly a third of the population of over twelve hundred souls were killed. The only reason that the Virginia Settlement survived was due to the action of one man. His name was Richard Pace who lived at Pace's Paines on the south side of the James River immediately opposite Jamestown Fort. This is his story.

CONTENTS

Introduction

Chapter 1:	The Thirteenth Mayde	5
Chapter 2:	I'm Coming Virginia	27
Chapter 3:	The Ancient Planters' Witness Statement	53
Chapter 4:	The Virginia Company and the Valley of Death	73
Chapter 5:	Richard's Return and the Marmaduke Maydes	101
Chapter 6:	Massacre – 'Trust is the Mother of Deceit'	115
Chapter 7:	Aftermath and Rebirth	135
Appendix	Author's Note Stepney Rediscovered	149

Introduction

This is the story of the first settlement in North America at Jamestown, Virginia. It is distinctive from other accounts because it is told here through the life history of one man. It is a cradle to the grave account of the life and times of the carpenter Richard Pace and his wife Isabel Smythe who dwelt at their holding called Pace's Paines on the south side of the River James immediately opposite Jamestown. But although this is the story of the life of one man the narrative covers the issues of the consequences of colonial seizure of territory, of class, leadership and man's struggle with nature. Until he had turned thirty Richard had a fairly conventional life, living and working as a carpenter close to the City of London. In the last ten years of his life he and Isabel made three hazardous voyages across the Atlantic in a tiny ship, endured great hardship and witnessed much sickness and death. Yet he survived long enough to know that his family faced the prospect of a prosperous future. But without his decisive action to row the three miles across the River James in the dead of night on March 22 1621/22 to warn of an imminent attack by the indigenous Indians on the stronghold and population centre at Jamestown, it is unlikely that the colony would have survived.

Richard Pace was born to a yeoman family in Kingston-on-Thames in 1580. Soon after his birth the family moved to nearby Farnham and it was here that Richard's only brother Thomas was born in 1584. His father, also named Richard, died suddenly in 1587 and his mother Anne Browne sought sanctuary with her sister Joanne Browne who had married a Stepney carpenter John Clawson. The Clawson's only son Robert was almost the same age as Richard and so the cousins must have grown up together. Richard chose to become a carpenter and so it was his younger brother Thomas who inherited the family holding at Kingston when in 1611 their relative John Pace died without male issue. Richard married his sweetheart Isabel Smythe at St Dunstan's Church, Stepney, in October 1608 and soon they set off for Virginia with high hopes of a better life. Charges for their passage were paid by the Virginia Company so Richard must have been contracted as an indentured waged artisan and would have been granted a few acres of land for his own use. It is possible that Richard and Isabel took passage on the *'Sea Venture'* that was wrecked on Bermuda and their son George may have been born on the Island and named after Admiral Sir George Somers.

The early years of the Settlement 1607-1610 had been a disaster for by the spring of 1610 the early emigrants had been reduced to a group of 60 emaciated survivors after the 'great starving time'. Supplies arrived in the nick of time to prevent abandonment. Richard would have known Pocahontas, daughter of the Powhatan chief, who was later held as a hostage in Jamestown, renamed Rebecca and converted to Christianity. Around this time John Rolfe had begun developing a new strain of tobacco and in 1614 helped to establish a more amicable relationship with the local Indians by marrying Pocahontas at Jamestown church. Richard and Isabel had very probably known John Rolfe in England and they were almost certain to have attended the wedding.

Richard and Isabel, with their young son George, must have endured a terrible time during the early years of their time in Jamestown. It would not have taken them long to realize that they had been duped and that the chance Richard had taken in selling his skills to the Virginia Company in the hope of a better life had not paid off. The structure and inequality of the 350 or so strong community and the nature of the regime must have quickly become very clear. The top third of the group were made up of the nobility and gentlemen and, supported by the military, they exercised a despotic rule over the others. From the remaining fraction the waged artisans provided the skill base and the indentured labourers the muscle power, but both were driven remorselessly and treated similarly as virtual slaves. Gross unfairness was a generally accepted characteristic of the feudal system in England but this settlement was a tiny isolated unit in a faraway land where all faced ever present danger and risk. The extreme unfairness must have been all too apparent and in the face of no general sense of purpose and no obvious chance of quick riches, lethargy would have set in among the workforce. In such a situation trust in the leadership collapses, energy is sapped and workers shirk as much as they dare.

After four or five years of hardship and drudgery the life chances of Richard and Isabel were seemingly offered the opportunity to be about turned. By 1616 the Virginia Company was on the road to nowhere and unable to pay a cash dividend but it could and did exploit its vast land bank. The Company eventually realised that the only hope of making money lay in developing a commercial tobacco crop to meet a burgeoning home demand. In order to increase the scale of production the colony required a large number of motivated workers to farm the labour intensive tobacco crop. So the Virginia Company engaged on a reckless expansion programme which inevitably meant the colony would hurtle into a confrontation with the local Indians. The first stage was to distribute land as a dividend to the shareholder Adventurers and as a grant to the 150 'Ancient Planters' – the early pre-1616 settlers still left alive. Richard and Isabel were granted 100 acres each though this was not finally registered until December 1620. In addition, in 1618 the 'head right' system was introduced which used land as an inducement to encourage individuals to recruit, finance and transport new emigrants for a payment of 50 acres a head. Richard and Isabel took advantage of this offer and returned to England in 1621 to recruit six workers. His niece Ursula Clawson sailed back with them. Much of the recruitment by others was often indiscriminate. Waifs and stray children were rounded up with the assistance of the Mayor of London and sentences of convicted felons were commuted on condition of immediate transportation. In 1621 a hundred 'maydes' from respectable families were induced to make the voyage for marriage and the first twelve returned with Richard on the *Marmaduke*.

The Virginia Company was careless as to the fate of recruited emigrants. Many of these new settlers were regarded not for their suitability for a pioneer life but for their potential for generating profit – a disposable commodity of throwaway people. The mortality was truly awful. But the consequence of the expansion was that the grant of rights to land led to a land grab and the seizure of territory without any negotiation or respect for the indigenous population. The local people living around Jamestown were Powhatan Indians living in Chiefdoms subject to the ruler Powhatan. The survival of the early settlers relied on trading with the Indians though intermittent violence was a feature of this early period. The crisis came in 1621/22 when the Powhatan Confederacy responded with a carefully planned sneak attack and the murder of around a third of the 1200 settler population. Richard Pace had learned of impending attack and warned Jamestown in time and so prevented a much higher death rate. Richard died almost exactly a year after the massacre having just returned to his new plantation. Shortly after his death Isabel married their neighbour William Perry and together with her son George, they prospered in Virginia.

As a background framework for the following account it may be helpful to consider the period from 1607 – 1624 to be divided into five distinct phases:

1. All that Glitters is not Gold. 1607-1610

The early voyages up to and including De La Warr's 4th Supply in May 1610, although suffused with statements of noble intentions such as securing territory for the King and christianising the savages, were primarily about finding El Dorado. The mind-set model was probably based on the success of the Spanish conquistadors who had shipped vast treasure home. Those who subscribed to the Virginia Company were specifically promised a share of any gold and precious stones that were found. On arrival in 1610 De La Warr's early initiative was to close the two main outlying forts and consolidate the 300 or so colonists at Jamestown before setting out with 100 of them to seek gold in the hills. It was only because his handful of mining engineers were invited to dine by the indigenous locals and promptly murdered that the expedition never got off the ground. Samples of soil sent back to England proved to be fool's gold dirt.

2. The Big Stick. 1610-1616

After the failure of the huge nine vessel 3rd Supply voyage of May 1609, the loss of the flagship *'Sea venture'* together with much of the equipment and stores and the dreadful mortality during the 'starving time' for those who did arrive at Jamestown, there was a change of approach by the Virginia Company government. The militaristic order which had been in force during the earlier period was transformed to a harsher, brutal despotic rule which had little regard for the rights of the settlers or the obligations of the Company. Under the leadership of Sir Thomas Gates, and particularly Sir Thomas Dale, the policy priority was switched from a focus on securing riches and gold to the development of a more appropriate infrastructure and healthier location about fifty miles upriver at Henrico City, and later the adjacent Bermuda Hundred. An associated tactic was to engage in almost continuous conflict with the Native Americans.

The problem with the development initiative was that it required significant investment and yet yielded insufficient short term profit to satisfy the investors who grew increasingly restless back in England. By 1614, when Sir Thomas Gates left, the Company was already recognised to be in deep trouble and by 1616 when Sir Thomas Dale departed it must have been clear that it was close to collapse. The only good news during the period was the development of a suitably commercial strain of tobacco by John Rolfe and a peaceable settlement of the 1st Powhatan War through the marriage of Pocahontas and John Rolfe in 1614.

3. The Juicy Carrot. 1616-1621

The tipping point came in 1616 when without any profits or sufficient manpower to exploit the potential of the tobacco crop, the Virginia Company faced ruin so it was decided to exploit the only asset they possessed – the vast land mass of Virginia. Two key initiatives aimed to solve both of these problems were:

a) To grant Shareholders and the Ancient Planters 100 acres each for their 'personal adventure'.
b) To pay 50 acres to any person who financed each new settler.

The transformation was remarkable for the grant of personal ownership encouraged both a release of entrepreneurship and a considerable flow of new workers to the Plantations. The situation was further improved by the appointment of Sir George Yeardley as the temporary Governor who provided sensible and considerate leadership, and eventually, formal confirmation by the Virginia Company of the land grant. The Ancient Planters' reports of this time are verging on the euphoric or even ecstatic, citing excellent harvests, peace prosperity and goodwill among colonists who offered hospitality to each other whenever they travelled abroad.

4. Death in Paradise. 1621-1623

Not so good however was the terrible mortality among the new arrivals now pouring in under the incentive scheme. Secondly the Ancient Planters showed a total disrespect for the rights of the Native Americans by grabbing their newly acquired Dividend Land without consultation or agreement. This brought an end to the seemingly benign and tranquil relationship with the Powhatan Confederacy and precipitated the shock massacre of a third of the settler population on the morning of Friday March 22nd 1621/22. The failure of political insight came home to roost and the eighty plantations were reduced to eight as the cowed and frightened people abandoned their homes and sought sanctuary from further attack in the security of Jamestown. The harvest was predictably meagre that year and the colony endured considerable hardship. The settlers eventually began to consolidate their position and began to attack the Powhatan Confederacy and drive the tribes out of the area.

5. Rebirth. 1623-1624

The most significant changes which ensured the survival and prosperity of the colony were the introduction of structures that allowed local representation, the establishment of property rights and the proper operation of the rule of law to secure justice and fairness. In the years to come Virginia grew and prospered. This is the consolidation period after the trauma of the massacre when there was a realisation of the need for vital institutional change and the introduction of democratic processes which brought about the necessary change to secure the prosperous future of Virginia.

PART I

ENGLAND

1580/81 – 1609/10

CHAPTER ONE

The Thirteenth Mayde

1

THIS CHAPTER IS all about discovering the family background of Richard Pace and his wife Isabel before they set out for Virginia. Seeking the family origins of Richard Pace was a challenging task and the reconstruction was a Sherlock Holmes job for there were four key revelatory signposts along the journey, each of which proved to be crucial. The search involved quite a few twists and turns and some perplexing issues but over time the situation began to unfold and a logical narrative has been constructed. Here the facts are presented and a case made but after over four hundred years absolute certainty is rarely possible. The conclusions have been made on a judgement based on the available evidence.

The Ancient Church of St Dunstan, Stepney

(Photograph Pace)

Members of the Pace Society of America, an organisation whose main objective is genealogical research into the history of the family, have over many years diligently searched for evidence of Richard's parental origins. Although a significant data base had been established and a number of possibilities suggested, no definitive factual evidence has been found. Given the widespread practice of naming an elder son after the father the obvious search clue was to look for a parent called Richard. Though there was a record of the marriage of Richard Pace to Isabel Smythe at St Dunstan's Stepney on 5th October 1608 further examination showed that although the Pace family were present in Stepney in the sixteenth century there was no apparent nomenclature tie up with Richard. Again throughout the sixteenth century there was a known concentration of the Pace family in an inverted triangle between the Rivers Severn and Avon with the apex at Gloucester and the base line on the Worcestershire/Shropshire border. Name associations in the north Worcestershire villages of Pershore, Defford, Berlingham, Ecklington and Brinklow, did appear promising but although numerous prospective family trees were constructed, the search was abandoned as fruitless.

The First Sign

The first signpost breakthrough from the impasse came from tackling the problem from a different angle. Richard and Isabel returned to England for a short time to recruit six new emigrants before returning in the *'Marmaduke'* on 12th July 1621. The ship's complement included twelve 'mayds' who were being sent out as potential wives for lonely colonists. Two separate manifests of this supercargo were made. On the first one there are twelve names. The later one shows that the ship must have called at Cowes, Isle of Wight in order to allow one girl to be substituted for another. But the vital clue to discovering Richard's origins was an entry showing a thirteenth mayde named Ursula Clawson.

Richard was charged a levy of tobacco to defray the expenses of Ursula's passage to Virginia and a copy of the bill was sent to Virginia on the same sailing and is mentioned in the accompanying letter from the Virginia Company to the Council:

> *The tobacco that shalbe due uppon the mariadge of these maides we desire Mr Pountis to receive and to return by the first; as also the little quantitie of Richard Pace the Copie of whose bill is returned*

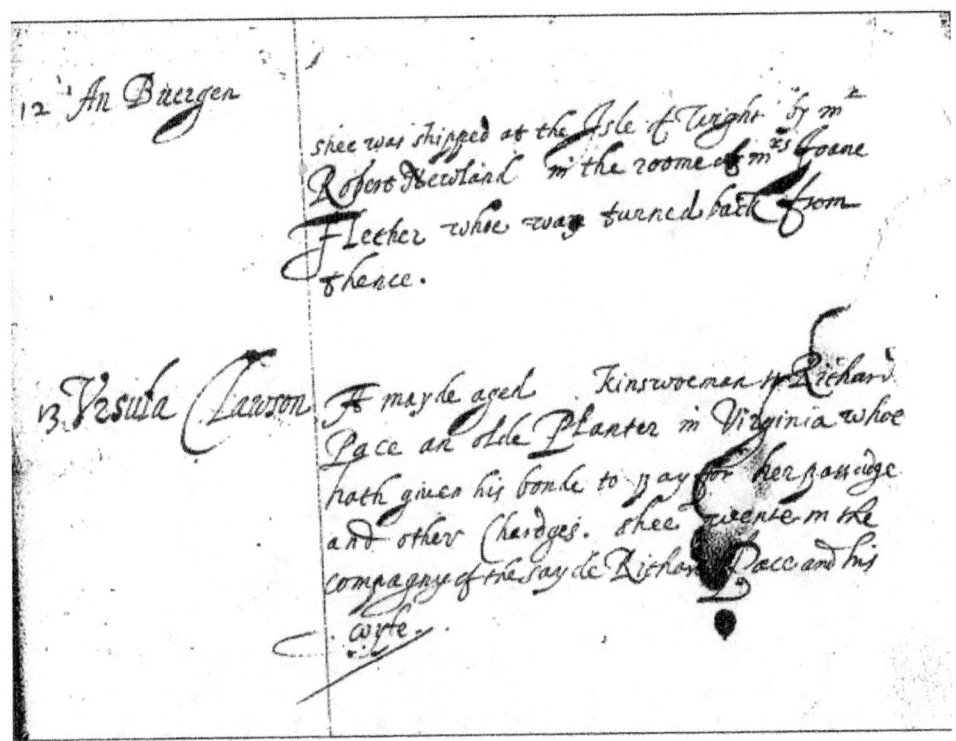

13. 'Ursula Clawson.
The mayde aged   . Kinswoeman to Richard Pace and olde planter in Virginia who hath given his bonde to pay her passadge and other Chardges Shee wente in the compagny of the sayde Richard Pace and his wife.'

Although as may be seen, a gap was left on the original document to record Ursula's age this information was unfortunately never put in. The fact that Ursula's name was only added at a later date may indicate a last minute rush and possibly that she was a local girl. But the key information is that she was a relative, possibly a blood relative, of Richard and not of Isabel. Fortunately Clawson is not a common name. A search of London records showed:

'Robert Clawson or Browne was baptised at All Hallows Church London Wall 11th August 1582'

Allhallows Church, London Wall (rebuilt 1767)
City of London Guild Church of the Worshipful Company of Carpenters

The parents are shown as John Clawson and Johan Browne. A subsequent search of a copy of the original record at Guildhall Library qualified the entry that Robert was the *'supposed son'* of John and Johan. The Clawson family have earlier records at All Hallows which show a common pattern over five generations of a father naming the first born son after his father. Further information that was to prove very important was that John Clawson was buried at St James Church, Clerkenwell on 12th September 1598. His widow very probably married a Richard Bigdenstone, again at All Hallows on 22nd June 1606. There is a subsequent record which shows that their son Robert

Clawson had a daughter Margaret who died in 1613 so he possibly married around 1605. Although no baptismal record has been found Robert may also have been the father of Richard Clawson and Ursula Clawson both of whom would have been born around the first decade of the seventeenth century.

A reconstructed family tree is shown below:

Key * = Event at All Hallows Church London Wall   **Event at St James Clerkenwell

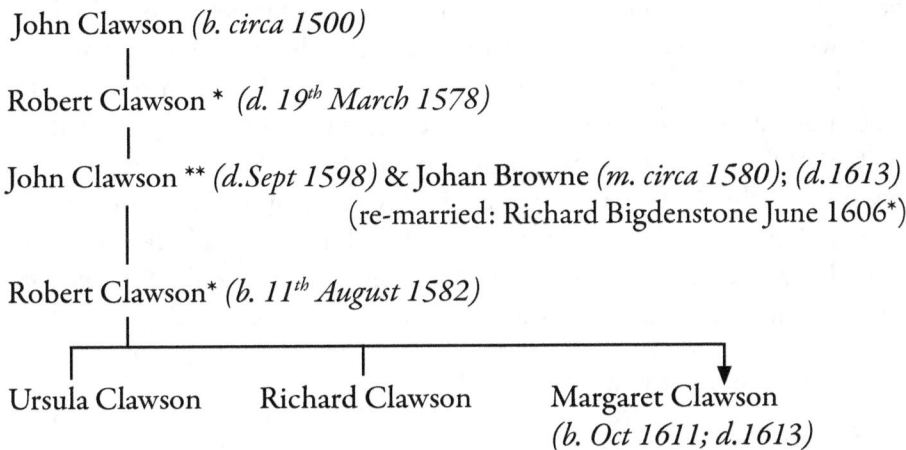

Further investigation revealed the significance of All Hallows Church when it was found to be the guild church closely associated with the Worshipful Company of Carpenters which has been based at London Wall since the first Hall was built in 1429. The Carpenters' Company has held its annual elections in the church for over 600 years. It was incorporated by Royal Charter as a City of London Livery Company in 1477 by King Edward IV to look after the welfare and interests of carpenters working in London and with the powers to regulate and maintain standards in the building trade in the City of London. Unfortunately the Registers of Apprentices only dates from 1654 and the freemen from 1673 but the Clawson family connection indicates a close connection with the carpentry trade.

This certainly moved the investigation forward but although Richard Pace and John Clawson were carpenters and Ursula Clawson was Richard's relative the nature of the family connection remained a mystery. But two key facts resulted from this information. Firstly it provided sufficient evidence that

Richard and Isabel who married in St Dunstan's Church in 1608 were in fact the couple who emigrated to Virginia. Secondly it was now clear that Richard and Isabel must have been settled Londoners and not recently from a more distant part of England. This allowed the search for the Richard's family origins to be considerably narrowed.

The Second Sign

The second breakthrough in finding Richard's family background came about because the search could now focus on a person who came from the London area, who quite possibly had a father called Richard. The fact that Richard called his son George was another vital clue to be borne in mind and later would prove to be of crucial importance. A review of the early English PACE RECORDS by the Pace Society of America data base drawn from parish registers in the 1500's, show a Pace family concentration in the County of Surrey about 10 miles from the City of London in two adjacent locations at Croydon and Kingston-on-Thames. The relevant entry on the Pace website for the birth of Richard of Jamestown:

> Kingston upon Thames
>
> Richard Payse
> Spouse Anne Browne
> Marriage 15th June 1579 Kingston on Thames, Surrey
>
> Richard Pace
> Christening 24th August 1580 Kingston on Thames, Surrey

There is an added note to the entry:-

> *'There is little doubt the above Richard Pace christened 24th August 1580 was the son of RICHARD PACE and ANNE BROWNE who were married in the same parish some 14 months previously'.*

> *'About four years later a RICHARD PACE shows as father in the christening of a Thomas pace in a nearby SURREY parish. It appears possible this THOMAS could be a brother to the RICHARD christened in 1580. A THOMAS PACE shows up later in KINGSTON parish as father of JOHN PACE christened 25th August 1616'.*

The entry covered the criteria for both location and the father's name of Richard. But possibly of greater significance was that it shows that Richard Pace married Anne Browne in August 1580. John Clawson was married to Johan Browne. The fact that their respective wives had the same name may show how Ursula Clawson was a kinswoman to Richard and Isabel. If they were sisters, then Robert Clawson was the first cousin of Richard Pace of Jamestown. It seems likely that Anne Browne and Johan were local girls for a parish record, which may be significant, shows that the previous year on 15th June 1579 a Joan Browne had married Owen Humphrey in the same Kingston church. Though no evidence has come to light it is possible that Owen died not long after his marriage and his pregnant widow married John Clawson who accepted the child named Robert as his own. The Pace and Browne family were well acquainted for in March 1606/7 Thomas Payce witnessed the proving of the will of William Browne a weaver at Kingston-on-Thames.

In the 16th century Richard's probable place of birth Kingston-on-Thames, was a significant, bustling market town close to the great palace of Hampton Court just across the river. The inhabitants of the town would have derived a good deal of trade from supplying goods and services to the palace particularly, when Queen Elizabeth and perhaps two thousand courtiers and followers were in residence. Kingston is about 10 miles upriver from Westminster which might take about 2 hours with a favourable tide. The Thames would have been a teeming traffic thoroughfare and it was far more convenient for travel than by road. Kingston-on-Thames was a place where justice was dispensed. Perhaps Richard's father was in the Market Place on the days that the records show that six vagabonds were hung and a female subjected to three immersions using the new ducking stool:

> *19th August 1572*
> *'On Tewsday being the xix day of this monthe of August Downing wife of (--)Downing gravemaker of this paryshe she was sett on a new cukking stole made of a grett hythe & so browgh aboute the market place to temes bridge And ther had III Duckinges over hed an eres becawse She was a common scolder and fyghter'*
>
> *8th September 1572*
> *'This day in this towne was kept the syce & sessions of gayle delyverye And was hanged vj persons and sertine taken for roges and vacaboundes and whypped abo the market place and brent in the eares on a munday'*

All the expected occupations are recorded such as butchers, bakers, carpenters, weavers, haberdashers, innkeepers, shoemakers, tailors, wheelwrights, fishermen, (eels and salmon were plentiful in the Thames), and husbandmen, (landless farm workers), and masons. Fights, drownings, murders and suicides are recorded. The law required death by plague to be identified in the parish register. Plague was a frequent visitor: 1563 – 78 deaths; 1593 – 81 deaths; 1603 – 86 deaths. Some of the royal officials and guards are mentioned as present in the town. The entry from the Early Pace Records recorded the birth of Thomas Pace at Farnham in 1584 and identified him as the likely son of Richard and Anne Browne whose marriage entry is shown. It was unclear why the first son assumed to be Richard of Jamestown was born at Kingston-on-Thames in 1580 while his brother was born at Farnham about 40 miles to the west until the third sign was discovered.

The Third Sign

The third signpost came from the microfiche records of the West Surrey Family History Society showing the death of Richard Pase at Farnham on 15th February 1587. This was a breakthrough for it implied that Richard's father was not the primogeniture son and had probably moved permanently to Farnham to seek work. His father possibly also lived in the town. This was a typical situation as second and subsequent sons often sought either an apprenticeship or waged work elsewhere and sometimes had to accept quite menial employment. If this entry does show the death of Richard's father, then his wife Anne Browne was widowed early in her marriage with two young sons aged 7 and 4 with no husband to support her. This would also explain why no other children are recorded as born to the couple. Neither son named his first born Richard and this may because they hardly remembered their father and chose to name their sons after more significant people in their lives.

The practice of transmitting land by primogeniture meant that the new head of the family, the elder son who embodied the family roots was attached to a particular place where the family had land. The Parish Records show that John Pace who died at Kingston in 1611 and Agnes who died there in 1608 are the only other possibly relevant Pace family entries so John Pace was probably the eldest son who inherited the land holdings at Kingston-on-Thames. The likeliest possibility is that John was the uncle of Richard of Jamestown. This position would possibly have included accepting responsibility for a member of the extended family who was in distress. Perhaps Anne Browne and her two young sons were offered refuge back in Kingston by John Pace. Her name does not appear again in the Farnham records.

Ancient Church of All Saints, Kingston-on-Thames where Richard was baptized 1580

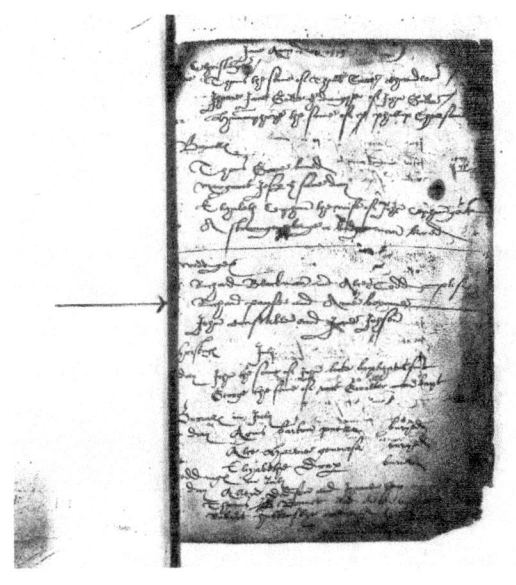

Parish Register Entry for the marriage of Richard Pace & Anne Browne

Another more likely scenario is that Anne Browne sought sanctuary either with other Pace family or with her sister Johan Browne and her husband John Clawson in the East End of London. Both boys may have initially moved together but it is probable that the family divided early and that young Thomas was brought up as the son of John Pace of Kingston-on-Thames, who appears to have had no male issue. This would explain Richard's apprenticeship as a carpenter, which he may have seen as infinitely preferable to working the land, and also his close friendship with his cousin Robert Clawson. It could also explain why Thomas son of Richard of Kingston called his first born son John. Thomas had several children and it is through the Will of this eldest son John that we know the inheritance that Richard would have got had he not become a carpenter:

The Family Tree of the Pace family with rites of passage at the ancient All Saints Parish Church at Kingston-on-Thames was reconstructed and evidence of the family background was gathered. The children of Thomas Pace were all baptised in Kingston-on-Thames which gives additional credence to the claim that he is the brother of Richard. In his Will of 1656 which has been accessed from the National Archives, John Pace is described as a yeoman farmer.

The land holdings appear to have passed to Richard of Jamestown's brother Thomas and thence to his eldest son John which would make John the nephew of Richard Pace of Jamestown. It gives a useful insight into the family background and indicates what Richard's inheritance might have been had he not become a carpenter. It appears that his inheritance would have been an owned small land holding of two acres at Walton-on-Thames and a larger holding at Kingston which John appears to have instructed to be sold five years after his death. There is a holding at Hooke in the Manor of Kingston-on-Thames, (this is today known as Hook and lies on the extreme SW of Kingston-on-Thames). This appears to be a copyhold tenancy and John instructs that the Reversion should be to his daughter after the death of Alice and Stephen Saunders the present incumbents. The overall picture of the Pace family is that they were long established in Surrey and came from a typical peasant farmer background. This meant that they would probably have owed service to the Lord of the Manor of Kingston in return for a copyhold tenancy of perhaps ten acres. (The Lords of the Manor of Kingston had recently included Elizabeth I and her chief advisor Lord Burghly).

The Thirteenth Mayde

The Family of Richard Payse

Key
 * Event at All Saints, Kingston-on-Thames
 ** Event at St Andrews Farnham

Note:
  Thomas Paeyse *(d. 11th Feb. 1545)**
  Richard Payce *(d. 7th Feb. 1593)***

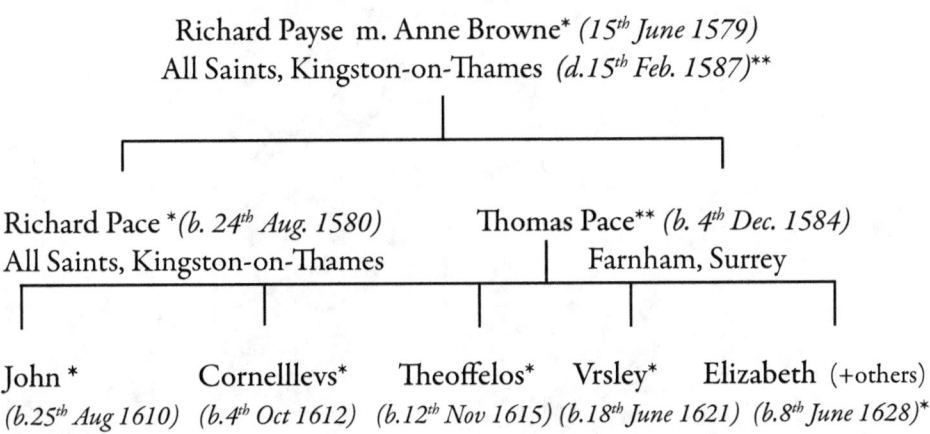

Richard Payse m. Anne Browne* *(15th June 1579)*
All Saints, Kingston-on-Thames *(d.15th Feb. 1587)***

Richard Pace **(b. 24th Aug. 1580)* Thomas Pace** *(b. 4th Dec. 1584)*
All Saints, Kingston-on-Thames   Farnham, Surrey

John * Cornelllevs* Theoffelos* Vrsley* Elizabeth (+others)
(b.25th Aug 1610) *(b.4th Oct 1612)* *(b.12th Nov 1615)* *(b.18th June 1621)* *(b.8th June 1628)**

On his death in 1656 John Pace is recorded as resident at Canbury Farm and the status of this land is less clear. This may well have been the property where Richard of Jamestown was born. Today Canbury is a district in the northern part of Kingston-on-Thames running down to border the River and not far from where Mrs Downing was ducked. The district takes its name from the historic manor which covered the area which included part of the town. John Rocque's map of 1746 shows the area comprising a patchwork of large fields transacted by a few roads, the principle north-south route being Canbury Lane. The manorial holdings included part of the open fields and buildings in the neighbouring Parish of Ham and which also lay within the Parish of Kingston-on-Thames. In 1652 it belonged to Arabella of Kent who would probably have been John's landlord.

The Ancient Church of St Andrew, Farnham where Richard's father was buried 1587

The Thirteenth Mayde

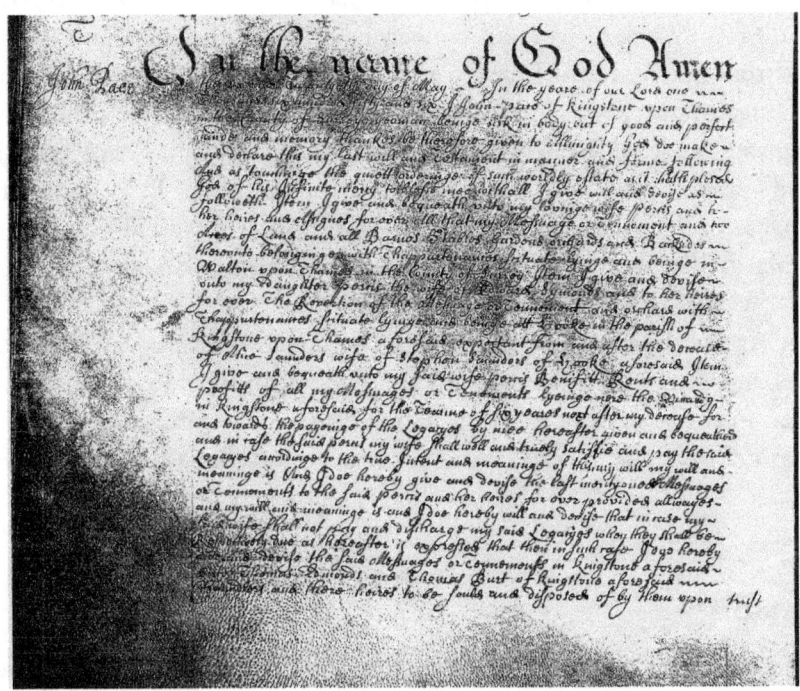

*Will of John Pace Kingston-on-Thames 1656
(nephew of Richard Pace of Jamestown)*

*The font at St Andrews, Farnham where
Richard's brother Thomas was baptised 1584*

The previous three key signposts that have proved helpful in discovering Richard's family origin were all concerned with hard data. The information showed that Richard and Isabel were certainly the couple who had married at St Dunstan's in 1608. The carpenter John Clawson, to who he may have been apprenticed, was Richard's uncle by marriage to his mother's sister Johan. Richard's parents were married at Kingston-on-Thames in 1579 and that as his father was not the primogeniture son, the family had moved to Farnham where his father died young in 1587, leaving his wife with two young children and no breadwinner. Richard was brought up in the East End of London by his mother Anne while his younger brother Thomas was raised in Kingston-on-Thames and by 1611 had inherited the family land holding. This process is depicted in the following diagram which also shows some of the Pace family resident at Clerkenwell whose importance becomes apparent under the Fourth Sign.

Scenario:

On the death of her husband in Farnham in 1587 Anne Browne seeks refuge with her sister Johan Browne in Stepney. Her eldest son Richard is possibly apprenticed to Johan's husband the carpenter John Clawson. Her younger son Thomas is adopted by John and Agnes Pace of Kingston-on-Thames.

Farnham	Kingston-on Thames	Clerkenwell	Stepney
Richard Payse & Anne Browne	John & Agnes Pace	Robert Pace	John Clawson & Johan Browne
	(No Male Issue)		
Richard Thomas		Salomon, Robert & Barbara	Robert

(A Richard Payce, recorded as dying at Farnham in 1593 may well have been Ann Browne's elderly father-in-law which would mean that that his son Richard pre-deceased him. Transposing his death with that of his son would make no significant difference to the subsequent account of Richard of Jamestown's early life).

These facts tell nothing about what sort of people Richard and Isabel were but it was the last of the four signs that, by providing vital soft data, allowed considerable insight into Richard and Isabel's character.

The Fourth Sign

The fourth signpost was a seemingly innocuous entry in the parish register entry of St James, Clerkenwell, which showed that the carpenter John Clawson was buried there in September 1598. This seemed strange when a funeral at possibly the Guild Church of the Carpenters Company might have been expected. In the event it proved a breakthrough to understanding Richard and Isabel's religious beliefs. Search showed that other Pace family worshipped and had rites of passage at St James. The two sons of Robert Pace, Salomen and Robert, who were very possibly related to Richard, were baptised there. A point of particular interest is a record of a marriage by Licence of Thomas Pase to Alice Old, St James Church, Clerkenwell, July 1609. The Old family feature in the Parish records of nearby St Giles, Cripplegate. It is possible that this is a record of the marriage of Richard's brother Thomas born at Farnham in 1584, the younger son of Richard Pace and Anne Browne. If so then this would establish a clear link between Kingston-on-Thames and St James, Clerkenwell. It is significant that the marriage is by Licence indicating that the couple came from another Parish. The timing of the marriage fits with the birth of the first son John born to Thomas of Kingston-on-Thames 25th August 1610.

But first an appreciation of the nature and intensity of the spiritual divide in the Protestant Church is necessary to understand the framework for Richard and Isabel's decisions. An era might be captured by a zeitgeist or spirit of the age such as the Renaissance or the Enlightenment. The zeitgeist for Richard and Isabel was the playing out of the consequences of the Protestant Reformation. Throughout their upbringing England was ablaze with religious fervour. A febrile atmosphere and antagonistic intensity was the backdrop to society and to everyday life. The Elizabethan settlement had resulted in two distinct and opposing strands of Protestant commitment. The increased dissemination of the English Bible and the urge to read the Holy Scriptures gave a big boost to literacy. In the 16th.C male literacy increased to 25% and female literacy from 1% to 10%. A growing individualism was implicit in the increasing importance that literacy played in developing religious awareness, for people could now find their own truth in the gospels. They could now read the revolutionary message preached by Jesus about the need to renounce wealth, power and violence and to set at liberty the poor and the oppressed. The St James, Clerkenwell Parish Register of Burials quite commonly shows the cause of death as 'died in the fields' – which implies that literally, sometimes people worked till they dropped. The common people could see only too plainly that the message of hope from Jesus was at odds with their everyday experience, and this realization spawned dissent and heresies, promoted cynicism about the clergy and led to traditional authority and practice being questioned. The religious settlement failed to satisfy the more radical evangelicals and this gave a political edge to their vehement opposition to what they perceived as the compromising doctrines of the official church – and so they attempted to down them.

The following notes capture the essence of the background situation in Elizabethan England. They are from *The English and Their History*, (Allen Lane 2014), by Professor Robert Tombs:

'The truly serious issue was religion, the focus of cultural, personal and political life to an unparalleled extent. In the wake of the Reformation and the impact of the English Bible most people felt more intensely about religion than all but the most fervent minority today ... religion had an inescapable social and political impact.'

'The Elizabethan church was the least Protestant of all the reformed churches ... diverse elements were left alone if they did not make trouble'

Elizabeth had no sympathy with hardliners in either camp and considered 'what they disputed about were trifles. The Queen did insist that people should obey the law at least by a minimal amount of apparent conformity and this included regular attendance at church'

'The ligament of society was the Parish – the spiritual embodiment of the community – on average 500-600 people of whom a quarter were adult males. The Parish was meant to reflect the political and social hierarchy with clergy appointed by the Crown, local landowners or Bishops with respectable men taking turns as Parish officers'

'Puritanism was a source of disruption – the godly who wished in a variety of ways to purify the Church from every relic of 'superstition'.

They were overwhelmingly Calvinists believing in Predestination – that God had chosen an 'elect' for their salvation. They believed that God controlled everything that happened and sought evidence in divine favour or disfavour in every event. Believers were impelled to convert others and reassure themselves by their godly zeal that they were among the elect – this involved sober dress, perhaps christening children with names like Hezekiah or Patience, preaching the word and combating sin'.

'The godly were far from popular for their aims involved the suppression of traditional games and pastimes. By 1620 they included many of the socially influential ... and appealed to people of all regions, women, the lesser gentry, smaller merchants, and educated craftsmen.'

'Across the country there were angry disputes as to whether the clergy should wear surplices and what the communion table should be called. The anger came from what these represented. What was at stake in these religious tussles were power struggles, social upheavals and cultural intrusions comparable with the impact of Islamism today'

In summary, these notes by Professor Tombs indicate the extent of compromise and equivocation surrounding the Elizabethan religious settlement and provide a helpful background against which Richard and Isabel took their decision to leave for Virginia.

A startling discovery which clarified a great deal was the finding that probably from its first beginnings in the early days of the Reformation, St James appears to have rejected the pomp and circumstance of a more traditional Anglican liturgy. St James lies about one mile north of St Paul's Cathedral in Clerkenwell. Here the streets still follow the format laid down in medieval times with many twisting streets and alleys.

From 1000 – 1539 this site was the Nunnery of St Mary. After the Reformation the former nunnery building acquired a second dedication to St James and the founding fathers at the time of the re-dedication appear to have slanted the church organisation and the form of worship to a more radical orientation. For instance by 1656 the parishioners had obtained the right to appoint their own vicar and these elections continued until the twentieth century. Today St James still maintains its Low Church principles, aiming to teach faithfully and accurately the word of the Bible unfettered by the distractions of tradition or ceremony. The values and principles of the founding fathers of St James, even today are reflected in the continued lack of the use of robes, alter cloths candles and liturgy in the church services, all signs of a more radical puritan orientation.

During the counter-reformation under Mary I 1549-56 there were around 400 burnings for heresy, of which 85% were in London. Many of these were at Smithfield adjacent to Clerkenwell and this would still have been a vibrant folk memory. A more radical religious attitude was a logical outcome of the Reformation as ordinary people were more able to make up their own minds on spiritual matters. This sense of independence and commitment would have been reinforced by some having witnessed horrific suffering for personal belief, quite possibly inflicted on people known to them and whose creed they shared. The evidence strongly suggests Richard and Isabel were linked to the radical evangelist wing. Shaped by these influences they would have been a serious minded independent, articulate and literate couple. The joy in their lives would come from sharing their beliefs with others of a like mind and behaving in a manner that might qualify them for eventual entry into the Kingdom of God. They would not only have held a shared doctrine with others of a similar persuasion but would almost certainly have shared strong personal relationships and lived locally among them. They may well have been Calvinists, believing in predestination and their joint decisions would have been taken in harmony with their beliefs. The names of dissenting evangelical Puritans would still be found in parish records even though they would not normally have worshipped at the local parish church, except to the extent that church attendance was enforced by law. The only way to record births and marriages was by baptism and marriage ceremony in your Anglican parish church. Similarly the only way to get a Christian burial was in your parish churchyard. There is evidence of dual church membership among the Pace family of Clerkenwell, where perhaps they found the Anglican low church ritual at St James church an acceptable compromise. Jamestown was an Anglican foundation where there would be a required obeisance to religious orthodoxy. Richard and Isabel would have had to outwardly comply with the official form of service and this suggests that religious motives were not the primary reason why they emigrated.

A small number of families are prominent in the St James records. This is not an unusual feature in parish registers for it is an inevitable consequence of a historically static society. Records also typically may show a creeping migration to the immediate area as younger sons and daughters found marriage partners or employment in nearby towns. These family groups were often part of a dispersed extended family with relationships dating way back to previous centuries with linkages maintained by return visits to the home village on Feast days or at harvest time. The St James Parish Register records twins named Richard and Isabelle baptized on December 11th 1610, the children of a George Davison. He may have been a very close friend of Richard and Isabel Pace and possibly the names of his latest offspring reflected this. The Browne and Smythe family names are very numerous in Clerkenwell and the Pace family was also significantly established. The record also has entries with variations on the name Rolfe. It certainly would not have been a coincidence that after being raised in England, in 1634 Thomas Rolfe decided to marry at St James, Clerkenwell before he returned to Virginia to claim his inheritance. These parishioners were probably on occasions joined by friends and other family who shared their religious persuasion coming up or down river from Stepney and Kingston, on a convenient tide by way of the main River Thames thoroughfare. These committed people socialized together and married each other. And sometimes they migrated together.

The Old Church of St James Clerkenwell

CHAPTER TWO

'I'm Coming Virginia'

2

THIS CHAPTER IS about why and when Richard and Isabel set out for Virginia. It is likely that there was not one single issue that dominated their decision to emigrate but rather it was a combination of things that they considered. As to when they departed there has been much conjecture as to the date they left England. The evidence suggests that it is more than likely that given the possible window of 1609-1616, they left earlier rather than later. But a startling finding has been that the evidence suggests that there is a strong likelihood that they sailed in 1609 on the '*Sea Venture*' and were wrecked on Bermuda and that this is possibly where their son George was born. This claim is controversial but the facts presented here support this conclusion. So because of this proposition there is a detailed description given here of the transatlantic voyage and subsequent ship wreck of the flag ship on Bermuda that is believed to have been a template for Shakespeare's '*The Tempest*'. Their decision making process to sail to Virginia and the subsequent shipwreck, is followed by a description of the situation Richard and Isabel might have confronted on their arrival in Jamestown a year later than they had originally expected.

Firstly it is necessary to reflect on the reasons why they decided to emigrate. It is stating the obvious to say that Richard and Isabel took a conscious decision to sail to Virginia. But movement requires a force to overcome a position of inertia and shift a body in a different direction and so it is worth thinking about how the couple came to take their decision to leave England. Sometimes a migrant would have had no say in the matter as would have been the case of the waifs and strays swept up from the London streets or the felons committed to transportation to the colony, but there is no reason to doubt that the departure of Richard and Isabel was other than a free will choice.

Perhaps the couple started with a review of their existing position. There were four main pillars which restricted the personal freedom of a person of Richard's status in early seventeenth century England. A yeoman or small tenant farmer who worked on the land was subject to the feudal system with its long established precedent conventions and community feudal farming practices governed by the Manor Court that allowed little discretionary initiative. Those artisans such as Richard, who worked in a trade in the town or city were subject to the irksome rules and petty restrictions of the Guild. Civil life was under the rule of an absolute monarch who governed by divine right supported by an aristocratic retinue. Spiritual life was directed by an autocratic theocracy which required regular attendance and, at the least, an outward acceptance of the doctrines and rituals of the Anglican Church. It is probable, given that Richard was almost certainly a literate and sober artisan, that the decision was unlikely to have been taken off the top of the head in a frenzy of enthusiasm caught up in the excitement of the times, or conversely by a couple seeking to escape and find refuge away from a troubled land. Given that emigration would mean they would embark on a dangerous, and possibly a foolhardy voyage, it was not a decision that they would have taken lightly. Far more likely, given what is known about their later destiny, it would have been a decision taken together by a couple who were very close with a deep understanding of their joint strengths and weaknesses and also an awareness of their potential. The decision process would likely to have been an interplay between the deeply thought through facts and spiritual considerations. It was a decision surely undertaken after much prayer and consultation and under what they believed, was the certain protection of the Almighty. If Richard had reviewed his life in retrospect he would have been able to identify three key decisions that had a decisive influence on what happened to him.

The first of these decisions was his choice to become a carpenter, rather than accept the role of the primogeniture son. We know fairly precisely what this latter course would have meant, for it is possible to trace the pattern of Thomas his younger brother who did inherit the family lands. The 1656 Will of his eldest son John who was Richard's nephew, an extract of which was shown earlier, details the three family land holdings in Surrey. As a yeoman farmer Richard would almost certainly have remained in Kingston-on-Thames, married a local girl and probably have had several children. He would probably have enjoyed a reasonably uneventful and relatively prosperous life. It is difficult to imagine a more different life from the one he actually experienced after his marriage in 1608, had he followed a more traditional path.

The second key choice was his marriage to Isabel Smythe. The St Dunstan's Marriage Register records:

Richard Pace of Wappinge Walle Carpenter and Isabell Smythe of the same were married on the 5th day of October 1608

Although this account is focused on Richard it is more appropriate to regard Richard and his wife as a couple who shared the achievements and the pain of the events recounted here. Isabel was almost certainly raised in Stepney but her parentage has not been firmly established. On the evidence of a Virginia Court document of 1628 which stated she was then forty, it is probable that she was born in 1588. If so, then if she married in her baptismal church of St Dunstan's, Stepney, then the record of an unnamed female child of William Smythe being christened there on 7th August 1588 might well be her. Isabel must have been a remarkable person whose strength and support for her husband and young son enabled all three to survive the harsh times they endured together. Her religious beliefs would have contributed to her strength of character and it is probable that the couple's spiritual equality was matched by an intellectual equality. It is impossible to know just how influential she was but she was not a passive figure and should certainly be regarded as at least Richard's equal. She emerges from this story as an extraordinary individual.

The third key influence was the decision to enter into a contract with the Virginia Company to settle in Jamestown. In retrospect this proved to be a vow of 'till death do us part'. Richard and the Virginia Company became engaged in a sort of dance to the death and perished within a year or so of each other. And although this eventually proved to be the end of the line for Richard, it proved to be a new beginning for Isabel and George, his only son and heir.

The motivational impulse to emigrate may have been stimulated by an earthly or spiritual dissatisfaction with their current situation, or perhaps it was based on the recognition of a seemingly glowing opportunity, or a combination of all these factors.

Alternatively it may have been a response to a perceived divine instruction, no doubt with affirmative support from friends who shared their religious persuasion. If the decision came down to a choice between earthly and spiritual considerations, a possible weakness in determining any possible earthly benefits derived from voyaging to Virginia was whether it could be arrived at on the basis of sufficient information. Inevitably in a pioneer situation where few have gone before there would not be a body of reliable reference data, though there was certainly some overtly flattering propaganda to mislead. Subsequently, as the stories filtered back as to actual conditions in Virginia the level of uncertainty would have morphed into a more calculable risk for later settlers. Negative vibes from the colony seemingly did not motivate potential new recruits for settler applications appear to have fallen off significantly after 1610/11. But in 1608/9 when Richard and Isabel probably finally decided to emigrate, they did so in a state of great uncertainty. They would have had little inkling as to what to expect. Given the lack of information their decision to go must, to a considerable extent, have been an act of faith – and in that lay the possibility that they were making a huge mistake.

Given this background, a key factor taken into earthly benefit consideration must have been a rather vague hope of a chance of self betterment. For this young family self betterment may have meant a less constrained life, a freedom to explore, to imagine, to adventure and to live more according to their own beliefs and to escape the constraints imposed by Church and State. They may well have wanted more freedom to worship as they wished and more opportunity to improve their lives. But it is infinitely harder to improve one's lot if there is little cash. Richard and Isabel seem to have had little capital apart from perhaps some meagre savings. The evidence that they were later required to pay 'Quitrent' rent on their 200 acre dividend indicates that the charges for their passage were paid by the Virginia Company.

So all this suggests that that Richard and his wife could not afford to finance their voyage themselves and so resolved to commit their future to serve the Virginia Company. Under the terms of an Indenture Richard almost certainly signed, he would have undertaken to work as a waged artisan for a set number of years. The term would typically have varied to between 5 and 7 years. The Agreement would have included a plot of land of not more than 10 acres to build a house and grow their own food. There would also very probably have been an allowance of time, free from service to maintain and farm the plot. Isabel probably did not have an Indenture, but based on this later example, Richard's contract could have contained a commitment that his wife undertook some routine duties in support of the settlement:

> *'But it is agreed that if the wife of the saide ... Shall take paines and/be helpful in Cookinge washinge mendinge of Clothes and other housewifely ther shalbe a reasonable allowance made unto her for her paynes ...'*

Richard and Isabel may well have developed a positive view of emigration from external sources but such momentous decisions are typically accompanied by an optimistic frame of mind. These were exciting times, for the heroic exploits of Sir Francis Drake, particularly against the hated Spanish, had roused great interest and excitement among the populace. Similarly the earlier ambitious attempt by Sir Walter Raleigh to establish a permanent North American settlement, to mine for gold and silver and to christianize the Indians had romantic overtones. Even though his initiative failed, the mysterious disappearance of the 'Lost Colony' may have fired the public imagination and influenced Richard and Isabel. There was certainly a lot of printed propaganda published at the time.

'News From Virginia' advertised an idyllic perspective:

'To such as to Virginia Do purpose to repaire
And when that they shall thither come Each man shall have his share
Day wages for the labourer And for his more content
A house and garden plot shall have'

In contrast the Virginia Company later ensured that any bad news was censored after the less than successful initiative to re-populate the colony in 1609. In 'Remembrances' sent to the Governor Lord Delaware in 1611, he is instructed to first establish a religion followed by good government and discipline with the proviso:

To suffer no letters to be written hither but such as shall be seene by my Lo. Or his counsell, for many rayling and reproachfull letters are sent hither which discourage all Adventurers and almost overthrow the action if the care and diligence of the counsell did not revive it'

In summary if Richard and Isabel set sail for any anticipated earthly benefit it would have been for a combination of reasons, together with an assessment of their present circumstances. There would have been a background of excitement generated by news and propaganda about the opportunities opening up in the New World. There was a seemingly attractive offer on the table from the Virginia Company for artisans such as Richard who had the required skills but little capital. The economic case for a move to Virginia may have been persuasive, for wage rates offered by the Company were seemingly more than twice what Richard would have been earning as a London carpenter. Personal reasons for the move could have included the influence of friends and family. There was also the issue of infant mortality. Richard and Isabel obviously intended to have a family. In the early seventeenth century around half of children were dead by the age of fifteen with a concentration among infants, and the couple would have taken into account the chances of survival of any child. Plague was a regular visitor and the symptoms were horrible: fever, a racing pulse and breathlessness followed by pain in the back and legs, thirst and stumbling. Buboes were hard swellings of a lymph gland called botches or plague sores which formed in the groin, armpit or neck and then ruptured. Finally speech would become difficult and victims would rave or suffer from delirium before succumbing to heart failure. It would have been awful to contemplate witnessing such a death, especially a death which hit the young disproportionately. Memories of a severe visitation of the plague in late 1606 when 600 Londoners died in October, would have reminded Richard and Isabel of the dangers inherent in London life.

However it is very possible that spiritual considerations were the most powerful factor determining why they decided to sail for the New World. This decision would have been of a different character from the purely practical advantages of emigration It is important to have insight into their frame of reference where religion was passionately and viscerally alive and so to comprehend the intensity of their religious gut feeling. The following analysis of the nature of their religious beliefs is speculative but is consistent with what is known of their background and place of worship. This was a time of great religious turmoil with the Thirty Years War struggle between Catholics and Protestant engulfing continental Europe. Richard and Isabel would certainly be unable to comprehend and cope with modern Western world's secular rationalisation of religious attitudes of belief in unbelief. The couple would have been in thrall to what they conceived as an omniscient and ever present deity. Richard and Isabel would have had a mind-set conditioned by awe and trepidation, constantly and single-mindedly trying to serve God in order to perhaps attain a state of grace in His scheme of things. Religious thought would have dominated their lives with their everyday talk infused with references, blessings and thanks for the providence and bounty of Almighty God. Living with the official Anglican Church meant that Richard and Isabel would have had to accept a degree of equivocation which was inherent in the Elizabethan religious settlement outlined earlier by Professor Robert Tombs. They may have been evangelical radicals but they were not absolutists for that would mean that any compromise would have been unacceptable. Perhaps the couple were wearied by the realisation of the near impossibility of securing any meaningful changes in the Elizabethan religious settlement. Even though the Low Church liturgy at St James, Clerkenwell probably allowed some leeway they would still have had to sit through the prescribed Anglican service which may have been an anathema to them. By an acceptance of the Elizabethan ambiguity they may have thought that they were not being true to themselves. Certainly in Virginia it would still be compulsory for them to attend an Anglican Service daily and to conform to the prescribed ritual. But a small wooden chapel in a distant land would be a very different proposition from the pomp and circumstance of an English Parish church and the associated Episcopalian authority. To evangelicals the church hierarchy would have represented secular power and an unnecessary and unacceptable barrier between themselves and God. So the opportunity for even a little more religious freedom may have played a part in convincing them to emigrate. Richard and Isabel may have felt that they were in an uncomfortable situation for they might have thought that they were not only betraying themselves by ignoring the reality of their compromise position, but also that they were betraying Almighty God. Living always with the shadow of purgatory in mind, this could mean eternal damnation. In order to resolve their dilemma they may have asked themselves questions such as: 'What does God require of me'? 'How can I best be of service to Him'? 'What is the meaning and purpose of life'?

A plausible explanation for Richard and Isabel's decision might be that it was primarily neither for earthly self advancement or enrichment nor for doctrinal spiritual reasons but because of a divine instruction. Very probably, they had prayed for guidance and believed that it was God's Will that they should sail to the New World and that this was his chosen purpose for them. It is possible to speculate that from at least the time of their marriage in October 1608, an intention to emigrate was central to their thoughts and was something they knew that they were going to do. It was not something they wished to happen or didn't wish to happen – the opportunity became part of their lives. After due consideration for them, the advisability of going to Virginia was no longer in doubt. They may have concluded that there was no need for an assessment of the balance of earthly advantage and disadvantage if this is what they believed God required of them.

The second key issue is the question of when they sailed. This is a list of the known voyages over the relevant period dating from the departure of the 3rd Supply in May 1609:

3rd Supply Voyage – Re-launch of Virginia Company after new Patent May 1609.

Sea Venture – flag ship wrecked Bermuda. (Capt Christopher Newport)
Swallow
Diamond
Unity
Lion
Blessing
Catch – Lost
Wicked – Lost
Virginia of the North Colony

Deliverance' and *Patience'* built on Bermuda – possibly 100 persons, not 140 arrived.

De La Warr 4th Supply Voyage May 1610 from Cowes Isle of Wight with 150 people

Hercules of Rye
Blessing
De La Warr

Sir Thomas Dale Voyage May 1611 – 300 men artisans and tradesmen

Starr
Prosperous
Elizabeth

Sir Thomas Gates Voyage June 1611 – 300 men

Trial
Sarah
Swan
+ 3 carvelles

Other arrivals identified by Ancient Planters:

Daynty October 1610 – small ship 12 men and 1 woman, some little provisions
John and Francis 1611 – small barque and a few men
Hercules 1611 with 30 people,
Sarah – small ship men wholly employed in trade
Treasurer 1613 – Capt. Samuel Argyll – 50 good men
Elizabeth 1613 – small store of provisions only, took Sir Thomas Gates home.

It is very probable that discounting the original 1607 arrival and the 1st and 2nd Supply voyages which were before their marriage in October 1608, Richard and Isabel sailed on one of the vessels listed. Given that De La Warr sailed from the Isle of Wight rather than London, that the major influx of 600 new settlers brought by Dale and Gates primarily aimed to provide a workforce, and that new arrivals dramatically tailed off thereafter again points to the 3rd Supply voyage as the most likely sailing that Richard and Isabel embarked on.

Another issue which might have influenced the timing of Richard and Isabel's departure was the question of when to start a family and the associated concern with the practicality of taking an infant child on a small crowded vessel. George their son was probably born in late 1609, around a year after they married. For devout Puritans marital sexual intercourse was primarily for procreation. It is even possible to speculate that apart from producing George the couple might have been celibate. There is a discernible pattern for Isabel had a child soon after marriage to Richard and only had her next and last child immediately after her second marriage to William Perry, fourteen years later in 1623/24. Similarly, the recently married, John Rolfe's first child was born on Bermuda in March 1610, so his wife would have been pregnant on her departure from England. His second child was born to Pocahontas in 1615, soon after their marriage in Jamestown a year earlier, and he again fathered a child soon after his third marriage to Jane Pierce. Contrast this with Richard's younger brother Thomas living in Kingston-on-Thames who produced children as regularly as clockwork for eighteen years until 1628.

It is likely that, given a deliberate family planning regime, Richard and Isabel would have either decided to sail before the birth or delayed sailing until George was more able to withstand a hazardous voyage. They would also have to coordinate with the infrequent voyages to Virginia. Richard and Isabel certainly had a strong inclination to go to the New World and given that the firm decision may well have been taken at the time of their marriage, it may well be the case that they decided to sail before their child was born. If they did not sail early then the birth of George in 1609 would have delayed matters and perhaps even given pause for thought about the wisdom of emigration because after 1610 Richard and Isabel may not have been so keen to go for news of the starvation and harsh conditions in Virginia would have filtered back and hardly given them reassurance.

These considerations suggest that the balance of probability is that, as there appears to have been no reason for delay, for Richard, with Isabel possibly in the early stages of pregnancy, may well have chosen to sail amid all the excitement and enthusiasm accompanying the departure of the great fleet of nine ships in June 1609. Given the survival rate of those of the 3rd Supply who arrived as planned in autumn 1609 and endured the starving time when the majority perished, then the implication is that Richard and Isabel sailed with John Rolfe and his wife, who certainly was in early pregnancy, on the flag ship 'Sea Venture'. Unfortunately only 52 of the 150 persons on this vessel have been identified. This total includes seven women and it is unlikely that there were more than around twenty females on board. If this is what happened then George must have been born in Bermuda or Jamestown. Certainly in spite of diligent search, no record of his baptism has been found in church registers in England. On the assumption that Isabel was not with child at the time of her marriage and given what may be inferred as to her character there would not have been too much uncertainty or procrastination about the decision to go to Virginia.

An alternative argument is that one pressing and pertinent reason that might well have caused a delay would be if Isabel had conceived immediately after her marriage. This would mean that she would have been in the last stages of pregnancy in May 1609 and may not have been prepared to risk such a hazardous voyage. This would imply that that Richard and Isabel sailed when George was older, on a later voyage probably after 1611/12 and certainly before 1616. It is therefore not possible to say with certainty when Richard and Isabel Pace left their life and friends in Stepney and set out on the perilous journey to the New World, but the balance of argument is still strongly for an earlier rather than a later departure and probably on the 'Sea Venture'.

There is a further piece of evidence supportive of an early arrival in Virginia. Later Muster information details the neighbours resident at Pace's-Paine after the allocation of land made in 1620 to the Ancient Planters who had arrived before the departure of Sir Thomas Dale in 1616. These included Francis Chapman, PhettiPlace Close and Thomas Gates, (not Sir Thomas Gates), who according to 1624/5 Muster information arrived aged respectively 10, 14, and 18, all on the *'Starr'* 1608/9, almost certainly as company servants. These were Richard's closest associates at Pace's-Paine and their early date of arrival suggests trust and a firm friendship developed over many years of enduring hardship together. Given the reported age of these three youngsters it is unlikely that they would all have survived the dreadful early days and the 'starving time'. The date recorded for their arrival at the muster is probably incorrect and it is more credible to conclude that they arrived on the *'Starr'* with Sir Thomas Dale in 1611. They were all much younger than Richard who was a skilled carpenter, and so they may have worked under his compassionate direction and forged ties of loyalty and respect for one another. By 1620 these Ancient Planters had progressed from servile status as youths to significant independent land owners. This enhanced status must have tended to erode the sharply drawn class lines between gentlemen and artisans. Ancient Planters and skilled craftsmen would now have been increasingly valued for their skills and their contribution to the colony rather than based on their family background. They may have decided to choose their Dividends of land in close proximity to one another for they had known each other from the early days of settlement.

In summary a review of the evidence suggests that Richard and Isabel would have been more likely to have sailed with the relatively huge 3rd Supply fleet in 1609 and if so, given the dreadful death rate of all those who sailed and arrived as scheduled in the summer of 1609, it is far more likely they would have been on board the *'Sea Venture'* with John Rolfe, Richard's associate and probable fellow Clerkenwell worshipper. They must both have been an overly driven and committed couple for they wouldn't have voyaged across a wild ocean to an uncertain future without a shared overriding sense of purpose and self belief. They must have been so full of hope but if they did but know it, they were sailing to a hell hole.

The Third Supply Voyage

Baron De La Warr had been appointed Governor of the Virginia Company, and the 3rd Supply mission consisting of 500-600 new settlers and a very significant fleet of nine ships set sail from London on June 2nd 1609. Baron De La Warr did not sail with the fleet and only actually arrived in Virginia a year later. The largest vessel of the fleet was the flagship *'Sea Venture'* 150 tons, Captain Christopher Newport, which carried the key leaders led by Sir

Thomas Gates the interim governor pending the arrival of Lord De La Warr. The Admiral of the Virginia Company, the experienced and respected Sir George Somers who had financed the fitting out of the *'Sea Venture',* was also on board. This 3rd Supply Voyage initiative was on a very different scale from other efforts at settlement and makes the earlier attempts seem like pinpricks. It represented a new beginning and was the first serious attempt to populate Virginia. The sheer size of such a significant venture which involved the organisation of the largest English civilian fleet ever assembled, the raising of new finance and the recruitment of such a large number of settlers would have been a major London event and created a great stir of excitement.

Seventeenth Century Transatlantic Sailing Routes. Anoldo di Arnoldi & Matteo Florini (1600) 'America'
(*Courtesy Library of Congress Geography & Maps Division*)

Whether or not Richard and Isabel sailed with the Third Supply voyage they would still have had to cross the Atlantic. The fleet of nine ships set out from London and by 2nd June 1609 had arrived at Plymouth Sound. The passage down the English Channel would no doubt have been regarded as a 'shakedown' cruise and an opportunity to check out the ships and equipment and to make good any deficiencies at Plymouth. The fleet then continued westward and finally departed from Falmouth on the 8th June on what turned out to be a most fateful voyage. The chart shows the three cornered Atlantic, three thousand miles across and a thousand fathoms deep, bounded by Europe and half of Africa on one side and by the vast continent of America on the other. But the chart tells you nothing of the strength and fury of the ocean, its moods, its violence, and its gentle balms. It is a place to be best avoided in small vulnerable sailing ships of under 150 tons. In summer it would be expected to be relatively benign and this would have been for that reason that the great majority of the early transatlantic voyages were made during the more clement months. Mariners had learnt from earlier voyages that they could save more than a week by taking a 'parallel' path across the Atlantic in order to pick up the south east trade winds. So the fleet would have sailed down the Bay of Biscay to the bulge of West Africa between latitude 27 and 28 degrees and then steered due west passing close to the Canary Islands.

Sailors would have been able to gauge their latitude well enough either by the length of the day, by a sun sight probably using a back staff, or from the altitude of known guide stars above the horizon. But longitude could only be determined by dead reckoning. This would have used elapsed time, compass course, hour glass and the technique of using a log thrown overboard to judge speed through the water to establish a very approximate position. The safest guide to longitude was an extremely long sounding lead line to determine shoal waters when the ship arrived off the Continental Shelf of North America. The nine ships stayed together until 23rd July. On St James Day Monday 24th July and just an estimated eight days out from Jamestown, they encountered a most dreadful storm of which William Strachey later gave a most vivid account. The hurricane force winds lasted for four days and three nights. On the first day as the gale strengthened and skies darkened the *'Sea Venture'* was forced to cast off the smaller vessel she was towing – probably *'The Catch'*. The vessel began to take on significant quantities of water and by the second night was only kept afloat by constant pumping by manning three pumping stations in the forecastle, the waist of the ship and aft. A round the clock rota was established involving 140 of the 150 persons on board indicating there were probably no more than a few women present. All took their fair share with Admiral Somers and Sir Thomas Gates spelling each other for hourly sessions pumping alongside a common labourer who was stripped to the waist. It was to be a rare occasion when any sort of equality was reported in the history of the Virginia Company. As the water rose five feet above the ballast the carpenters were sent below and made frantic efforts to find leaks and repair the caulking. All moveable equipment was jettisoned to lighten the vessel. William Strachey later complained that the loss of some personal effects had cost him dear. It must have been obvious to Admiral George Somers that he and the crew were fighting a losing battle. When on the third day the ship was 'pooped' he must have known that the end was close. This event meant that if someone on the afterdeck had glanced over their shoulder they would have looked up at a mountain of water perhaps thirty feet high. Without the burden of the seawater in her bilges the *'Sea Venture'* would have the buoyancy of a cork and lifted her stern, but weighed down and water logged the massive sea crashed over the decks. Strachey described this terrifying time:

> *'Once so huge a sea broke upon the poop and quarter upon us as it covered our ship from stern to stem like a garment or a vast cloud. It filled her brim full for awhile within from the hatches upto the sparre deck. The force and confluence of the water was so violent as it rushed and carried the Helmman from the Helm'*

On the fourth morning when all seemed lost it was the old sea dog Admiral George Somers who sighted the NE corner of the main island of Bermuda, possibly because he sensed the proximity of land and so was looking for it. He would have known that Bermuda was the only possible landfall that could have been in the vicinity. He may have perceived the water change from deep ocean green to a lighter grey, noted the long swell shorten to a shallow sea chop and even glimpsed the occasional seabird indicating the near presence of land. Whatever, George Somers was the man for the moment. He had distinguished himself afloat in 1595 in the Anglo/Spanish War and had later commanded several English naval ships between 1600 and 1602. He had been knighted in 1603 for his services. He was also very fortunate to have in support the very experienced sailor and former privateer Captain Christopher Newport who was on his fourth voyage to Virginia. Being the fine seaman he surely was,

Somers would have instantly summed up the situation and acted instinctively. The ship was sinking and time was of the essence. Assuming the *'Sea Venture'* arrived off St Catherine's Point on the NE tip of St Georges Island, from an examination of the chart it is obvious that his options were extremely limited and would have depended on the precise direction of the wind. The vessel was caught in a storm force gale on a lee shore. If the wind was from the NE the ship would inevitably have been driven onto the rocks on the northern coastline. If the wind was from the North or NW then Somers would have recognised the danger that she might be carried down the east coast and back out into the Atlantic. If it was a northerly gale then it would be crucial to round St Catherine's Point and founder on the south side of the promontory in order to disembark the ship's complement in a much more orderly fashion. His plan of action in this situation, in concert with Captain Christopher Newport, would have been to stand off the Point and then steer directly for the land. He might just have been able to influence how and where the ship would strike and this could make the difference between a handful of survivors or saving all on board. If it was a SE storm then the danger was she would be washed past the Island instead of ending up where she did, on the lee shore in Gates Bay.

Whatever the direction of the wind was, it must have been a close run thing. Somers would have immediately ordered sail to be hoisted to give the vessel more way and give steerage to bear up towards the land. Full of water the vessel would have been very sluggish and slow to respond to the helm and difficult to manoeuvre. Somers must have realised that if he could get the vessel to make some way through the water it might be possible to keep her head to land and sea and prevent her broaching as she struck bottom and being battered to pieces. The only blessing was that there would then be some shelter from the lee of the land and so the wind and sea would be abating. It appears that this is what happened and the site of the wreck on a reef off St Catherine's Beach in Gates Bay is indicative of purposeful action rather than luck. Somers would have ordered the boatswain into the chains to shout the lead soundings as they approached the island. Though the vessel normally perhaps had a draft of 12 feet, full of water she now would have had something like 18 feet below the waterline. If the ship sank even close to the shore many would drown for the ship's boats would not have been adequate to carry only a fraction of the 150 persons on board.

Somers would have known that the only hope was to ground the *'Sea Venture'* as close to the shore as possible and leave her upper works above sea level. The leadsman would have swung the lead full circle before casting ahead of the ship and quickly hauling the line perpendicular to sight the coloured bunting or marks and singing out the sounding. The first shout was thirteen fathoms then seven and finally four. Somers realised grounding on the rocks was the only option so he drove the ship as close in as possible before she hit bottom jammed between two reefs in about three fathoms. To help ensure that, rather than slipping back into deeper water and potential disaster she stuck, he would have ordered everyone to crowd down aft on the poop deck to get the ship's bow as high as possible and told them to brace themselves against the shock. She struck between two ridges of rock in on the north east side of Bermuda and just over half a mile from the shore. She must have come to rest with her topsides above water. It was at this point that he probably ordered one, or possibly two, anchors to be let go in order to hold her in position. It was truly a miraculous survival from certain drowning for few would have been able to swim. All passengers and crew were now able to be ferried to the shore in an orderly fashion, with no loss of life whatsoever. William Strachey later described the event:

> *'we were enforced to run her ashore as near the land as we could, which brought us within three quarters of a mile of the shore, and by the mercy of God unto us, making out our Boats we had ere night brought all our men women and children about the number of one hundred and fifty safe into the Island'*

The *'Sea Venture'* had come to rest off St George's Island, an uninhabited but bountiful refuge, which was able to support the unexpected arrival of 150 desperate souls. There was water, edible vegetation, trees for shipbuilding timber and even a colony of wild hogs left by previous visitors. But without the bounty accessible from the relatively intact wreck of their ship within fairly close proximity of the shore the castaways would surely have had to endure a much longer incarceration. Even if they had been discovered a passing vessel would have been unable to take such a large number on board and would have had to report their position for a later rescue. In the days following there would have been numerous visits out to the wreck and the crew would have dived again and again through the hatchways in an effort to pick the ship clean of any useful items before she split her sides. They would have made every effort to ransack the offshore hulk in order to salvage food, stores, seed corn, tools and nails to aid survival. But without being able to cannibalise essential equipment from the *'Sea Venture'* there could have been no early escape. The masts and spars, blocks and purchases running rigging, ropes and sailcloth were essential components to provide the wherewithal to construct the two barques *'Deliverance'* and *'Patience'* on which they escaped in time to make a fortuitous rendezvous with De La Warr's 4th Supply Voyage. The likelihood is that Richard Pace was one of only a handful of artisans with the necessary skills to build these vessels.

'Sea Venture' wrecked here. Crew landed in Gates Bay, St Georges Island, to the south side of St Catherine's Point

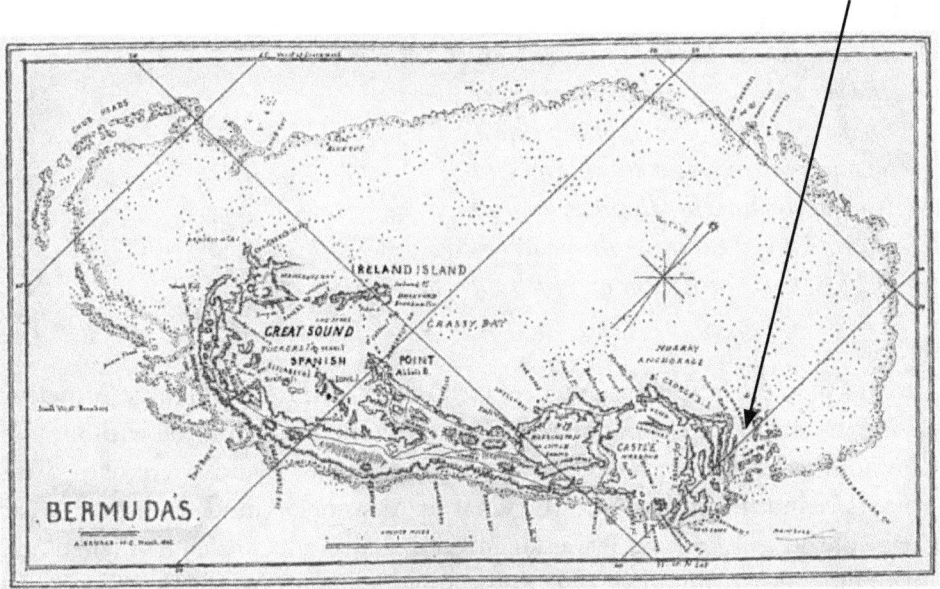

Bermuda Chart by Arthur Savage March 1862

Carried back from Jamestown by Thomas Gates and dated July 15 1610, Strachey's sensational account of these events was circulating in London by the autumn and many scholars believe that Shakespeare read it and used it as a template source and inspiration for *'The Tempest',* thought to have been written between 1610 and 1611. During this period of his life Shakespeare was a shadowy figure but he certainly had resided for some years in the vicinity of Clerkenwell, for when he appeared as a witness in a court case in 1612 he said he had known the family of his landlady Marie Mountjoy of Silver Street,

Cripplegate for 12 years. Cripplegate is close to Clerkenwell where the name Shakespeare can be found in the St James parish records. Whatever the truth of this, Shakespeare must have been caught up with the earlier general excitement with the departure of the nine ship fleet particularly as his patron the Earl of Southampton was a major player in the activities of the Virginia Company. And Shakespeare may well still have been living in the locality when the astounding news arrived of the deliverance of the *'Sea Venture'* and the almost incredible tale of the survival of all on board. Perhaps this stimulated his imagination to write a play about an island in the middle of the ocean where Prospero uses his magic powers to raise a storm which shipwrecks anyone who tries to visit. Strachey certainly had used the term 'unmerciful tempest' in his text:

> *The direful spectacle of the wreck which touched*
> *The very virtue of compassion in thee*
> *I have with such provision in mine art*
> *So safely ordered that there is no soul ...*
> *No, not so much perdition as an hair*
> *Betid to any creature in the vessel*
> *Which thou heard'st cry or which thou saw'est sink*
> *(Act 1 Scene 2 of 'The Tempest')*

Perhaps these lines were intended as a tribute to Admiral Sir George Somers in saving the entire ship's company of 150 men women and children without loss. The uninhabited and previously dreaded Bermudan Islands were named the Somers Islands in his honour. The Coat of Arms of Bermuda today carries an image of the *'Sea Venture'.* It was an amazing deliverance which had relied on a lot of luck. But Admiral George Somers had made his own luck. Whether or not he had been the decisive influence, he was lauded for a seemingly brilliant act of seamanship. The fact that the ship foundered in the position it did, not only allowed all on board to land safely on the island, but by salvaging a sizeable quantity of stores and equipment ensured that nearly everyone was eventually able to leave it.

What's in a Name?

There is some evidence that Richard and Isabel's son George may have been born on Bermuda. John Rolfe's wife was certainly just pregnant before they sailed for it was recorded that she gave birth in February 1610 on the island to a daughter named Bermuda, though sadly both of them died. Rolfe was classified as a gentleman and so this birth would likely to have been more noteworthy than a child born to the wife of Richard, a humble carpenter. There was certainly sufficient time for Isabel to become pregnant and to give birth on Bermuda, or possibly later in Jamestown. Richard's father who had married Anne Browne at Kingston-on-Thames, in August 1579, was almost certainly named after his father, who at the end of his life probably dwelt in Farnham. Richard of Jamestown was baptised at Kingston-on-Thames in June 1580 and was similarly named after his father. Richard and Isabel's son George would go on to name his first son Richard and this sequence was followed by the subsequent generation. This adds up to five sons named Richard with just a single exception and so raises the question as to why the mould was broken just once. The most likely reason that Richard of Jamestown did not call his son after his father is that he had died when he was very young so he hardly remembered him and possibly had little emotional attachment to the name.

Admiral Sir George Somers

But the reason why he called his son George has been an elephant in the room throughout this study. An apparently plausible and obvious reason is that he was possibly named in honour of Sir George Somers, the hero and saviour of the *'Sea Venture'* after whom Bermuda was renamed. If so, this must mean that Richard and Isabel must have been on the vessel and that George was born on the island or shortly after arrival in Virginia. This theory is consistent with a friendship with John Rolfe at St James Clerkenwell, where Richard's uncle John Clawson had been buried and where Pocahontas's. son Thomas eventually married. It also might explain the absence of finding a birth record for George in the English Parish Records.

There is a final tenuous clue as to which vessel Richard and Isabel sailed on. There is evidence that the network of the Pace and Davison families back in Clerkenwell were well known to each other. George Davison had married Jane Hucksley at St James church and twins Emaniel and Theodor were born in 1607. Perhaps in September 1610 George Davison heard the joyful news of the miraculous survival of the *'Sea Venture'* for he named his second set of twins Richard and Isabel, born on December 11th 1610.

Just as extraordinary as the safe landing of the entire complement of the *'Sea Venture'* safely on Bermuda was the fact that a year later the great majority of them sailed the further 800 miles to Virginia in two appropriately named vessels *'Patience'* and *'Deliverance'*. This was a truly remarkable achievement. It is clear that the hierarchy of social status was maintained on Bermuda among the castaways with the aristocrats and gentlemen merchant adventurers, no doubt supported by an armed militia, maintaining order and retaining their authority. Though there have been reports of tension between a Sir Thomas Gate's faction and a Sir George Somer's faction it is apparent that they must have exercised strong leadership and a sense of direction aimed at preventing anarchy. Bermuda was a very fortunate and hospitable place to be marooned with plentiful food and fresh water but the initiative to build two ships in which to escape would have added the essential ingredient of hope. In such a predicament situational leadership may be crucial and the plan would not have been feasible without the skills of the handful of shipwrights or carpenters. These few artisans must have enjoyed considerable kudos during the construction of the two ships and their contribution to survival may have been favourably remembered by the leadership in the years ahead in the colony. After ten months on the uninhabited Bermudas the castaways arrived in Jamestown on May 24th 1610 on the *'Deliverance'* and *'Patience'* but found the settlement in a terrible situation with few of those they had set out with so hopefully a year earlier still alive.

In William Strachey's words:

'all things son contrary to our expectations, so full of misery and misgovernment'

Sir Thomas Gates described the survivors as:

'lamentable to behowlde cryeinge owtt we are starved we are starved'

The fate of the fleet and the 500 or so fellow emigrants they had set off with so hopefully a year earlier was about as bad as it could be. On 4th August 1609 the *'Unity' 'Blessing', 'Lion' and 'Falcon'* of the original nine strong fleet had arrived safely in Jamestown, followed by the *'Diamond' and 'Swallow'* two weeks later. The *'Wicked'* and the *'Catch'* were presumed lost. The *'Diamond'* may have had significant disease on board. Of possibly 300 who had originally been carried in these six ships, when the castaways from Bermuda arrived they found only 60 emaciated survivors. And many of these survivors were found to be at death's door in the most pitiful condition and engaged in survival cannibalism. So a strange paradox is that the loss of the *'Sea Venture'* was the shipwreck that probably saved the Jamestown colony.

PART 2

VIRGINIA

1610/11 – 1622/23

CHAPTER THREE

The Ancient Planters' Witness Statement

3

THIS CHAPTER DESCRIBES the Jamestown settlement from first arrival in 1607 to just before the dissolution of the Virginia Company in 1624 as told by the settlers who experienced it. It is possible to read the music underpinning the words to gain insight in the attitudes of the settlers – what actually happened to them, what they thought about it, how they felt about it and what they did about it. It may perhaps be best regarded as collective evidence given by 149 Ancient Planters from the witness box covering the time that Richard and Isabel lived in Virginia as a married couple. It therefore provides a reasonably detailed background picture to allow an insight into the harsh conditions endured by them. It is in sharp contrast to a more abstract impersonal account which runs the danger of losing the essence of the content embodied in the choice of words for it allows access to the immediacy of the underlying emotions, the stoicism, and sense of hopelessness, despair, fear or tension that may have been present in the situation at the time. The experiential account therefore helps to develop an empathy and understanding of what it was like to live with the ever present danger of violent attack and the horror of regularly witnessing untimely and cruel death. It is drawn from a retrospective account of the first twelve years of the Colony authored by the Ancient Planters called *'A Brief Declaration of Virginia in the First 12 Years' (Feb 1623/24)*. Once the archaic format of this 400 year old document has been penetrated it is apparent that it is a beautifully written and concisely expressed account by the scribe Tristram Conyam. The document amounts to a record by those who actually lived through the events and is therefore more likely to be an authentic account though it is necessary to be wary of the possibility of bias as it was produced by a discrete group. The Declaration initially covers the period 1607 – 1609/10 before the arrival of Richard and Isabel and runs up until the time of her husband's death 1622/3. This summary gives an orientation and an account of the momentous last quarter of Richard's life. The Declaration rather resembles a 'round robin' critique of the 12 year stewardship of the

Virginia Company under Sir Thomas Smith from 1607 to 1619, but also comments on the period until 1622/23. There is biting criticism directed towards the Virginia Company Treasurer Sir Thomas Smith personally. As the Treasurer he would have been regarded as primarily responsible for the serial under-provisioning of the Supply voyages and particularly for the dire quality of some of the food sent out about which formal complaint had been made to London. Apart from this the Report adopts an almost philosophical approach to the horrendous events recorded.

The Report

Perhaps a more balanced judgement was made with a backward glance. During the immediate day to day lived experience it may have been harder to appreciate how life was changing for the worse. Richard had been dead about a year before the report was written but he would surely have wholeheartedly agreed with the views expressed by the 149 survivors of his peer group all of whom had also settled in Virginia before 1616. The Report is a measured account even though the major theme is the abject governance of the colony by the Virginia Company. Importantly the Report gives a long term perspective and offers a first-hand account and an objective view of what actually happened. What is surprising is that the tone of the report is more one of sadness than anger, more of a restrained diatribe than a vociferous condemnation. It is this aspect that gives confidence that the Declaration is a reasonably accurate unblinking registration of the actual events for the content speaks for itself. It is a most shocking document for it describes how, among others, skilled artisans such as Richard Pace, were systematically ground down, driven and abused in a pseudo slave regime. It details a catalogue of failure and foresight where self interest and avarice took precedence over concern, not only for the most basic welfare of settlers, but for life itself. The account of the sheer brutality of the militaristic control and punishment regime is astounding and even given the social mores of the time, it makes disturbing reading. Supporting evidence from numerous letters and accounts often confirms the statements of the Ancient Planters or indicates the frequency of hidden agendas of the participants in the drama and the manner in which presentation of the facts was sometimes distorted to curry favour with the recipient of the communication. Because of the crucial importance of the Declaration to aid understanding supported by the additional material, it has been relied on as a major analytical instrument.

The Ancient Planters' Witness Statement

Letters Patent had been granted to the Virginia Company of London on 10th April 1606. The Ancient Planters reported that the original Aims of the Colonisation were:

'The settlement in Virginia should be settled for the Glory of God and the propagation of the Gospel of Christ'

'The Conversion of the Savages'

'To the honour of His Majesty by first the enlarging of his territories and future enriching of his kingdom'

The first voyage with 104 settlers had arrived in early 1607 on the *'Godspeed' 'Susan Constant'* and *'Discovery'*. Captain John Smith later noted the vessels carried a very occupationally unbalanced group of only 82 persons consisting of 6 Council members, 48 gentlemen, 4 carpenters and 24 labourers. The Council sent an overly optimistic report to London dated June 22nd 1607:

'the land will flow with milk and honey if we are well supplied'

First Arrival in the James River
(Image courtesy of Gutenberg Project)

This would prove to be a forlorn hope for by September 1607 only 46 settlers were still alive. The Report goes on to describe the early years of this first Plantation of *'one hundredth persons'* under the mismanagement of Sir Thomas Smythe who aimed at nothing more than:

> *'particular gain to be raised out of the labour of such as both voluntarily adventured themselves or were otherwise sent over at the common charge'*

> *'the first plantation consisted of one hundredth persons so slendely provided for that before they had remained for half a yeare in the new colony they fell into extreme want not haveing anything lefte to susteine them save some ill conditioned barley'*

Christopher Newport returned again with the First Supply the following winter and arrived in the *'John and Francis'* in January 1608 – a voyage across the stormy North Atlantic – with 70 new settlers and totally inadequate supplies. There may have been a second ship that sailed at the same time but which did not arrive until April though this is not mentioned in the Declaration document. Newport returned yet again in the *'Mary Margaret'* with the Second Supply in September 1608.

> *'which arrived here ..... in her sixty persons most gentlemen – few or no tradesmen ... so meanly likewise were these furnished forth for victuals'*

The Muster of colony settlers dated 1624/25 frequently gives the name of the vessel the settler came on and the date of arrival. The names of nearly all the colonists arriving on each of the first three voyages are known. Some of this data is inconsistent with information set down by the Ancient Planters in the Declaration Report 1623/24. For instance the 1624/5 Muster shows John Proctor arriving on the *'Sea Venture'* in 1607 when he was actually on the 1609 voyage that was wrecked on Bermuda. It is possible, but very unlikely, that he had subsequently sailed back to England and was returning to Virginia. Reports have suggested that the reason the vessel had foundered was because it was newly built and the caulking between the planking was not yet firmly established. If so then the vessel had not been built in 1607. Similar doubt surrounds the listing of the arrival date of Richard's later neighbours at Pace's Paines, Francis Chapman and Pettiplace Close who are listed in the Muster as arriving on the '*Starr*' in 1608, a vessel not mentioned by the Ancient Planters.

The list below shows only those arrivals identified by the Ancient Planters in their Declaration Report, the contributors to which actually included a handful of those who were among the very first original settlers. The reliability of the information is dependent on a judgement as to the accuracy of the collective memory of the Ancient Planters who actually participated or otherwise probably witnessed the events. This is a summary of the information provided by the Ancient Planters as to vessels arriving in Virginia:

The Original 1606/07 Voyage: (Capt Christopher Newport)

God Speed
Pheonix
Susan Constant; 104 Settlers
Discovery

1st Supply Voyage 1607/8 (Capt Christopher Newport)

John and Francis; 70 Settlers

2nd Supply Voyage 1607/8 (Capt Christopher Newport)

Mary Margaret; 60 Persons – most gentlemen; few or no tradesmen

By November 1608 the surviving settlers from the first voyage and the 1st and 2nd Supply voyages were in desperate straits:

> '*want compelled us to employ our time abroad in trading with the Indians for Corn whereby for a time we partly relieved our necessities yet in May following, (1609/10), we were forced .... to disperse the whole Colony some among the Savadges but most to the Oyster banks where they lived upon Oyster for the space of nyne weeks which kind of feeding caused our skins to peele from head to foote as if we had been flayed*'

Back in London the early investors in the Virginia Company, increasingly unhappy with the accomplishments of the Jamestown colonists, sought a new charter which the King granted in May 1609. This 2nd Charter granted full authority and near dictatorial powers to the colony's Governor and allowed the Company to make its own laws and regulations subject only to their compatibility with English law. The New Patent initiated an attempt to colonise Virginia on a much larger scale through the 3rd Supply voyage under the newly appointed Governor De La Warr which left Falmouth on 8th June 1609.

> **3rd Supply Voyage** – Re-launch of Virginia Company after new Patent May 1609.
>
> Sea Venture – flag ship wrecked Bermuda. (Capt Christopher Newport)
> Swallow
> Diamond
> Unity
> Lion
> Blessing
> Catch – Lost
> Wicked – Lost
> Virginia of the North Colony
>
> 'Deliverance' and 'Patience' built on Bermuda – 100 Persons

The Report extract cited above describes the condition in which the 60 or so survivors from the first three voyages were found, when the six ships of De La Warr's 3rd Supply voyage arrived in July/August 1609 without the 'Sea Venture'. But for the new arrivals things were now going to get much worse. It is at this point that the Ancient Planters' record becomes directly relevant to the experience of Richard and Isabel Pace. A reasonable estimate would be that not many more than 200 settlers landed from the 3rd Supply to join the 60 or so semi-starving resident colonists. The vessels arriving were inadequately supplied and the newcomers soon shared the same difficulties of their fellow settlers.

The Powhatan Confederacy would not have welcomed them. They may well have been dismayed at the considerable number of new arrivals for there was already pressure on food after seven consecutive poor harvests. They would have been acutely aware of the threat posed by a well armed and disciplined force raiding their supplies when times got hard. And times did get much, much, harder. This latest attempt to settle Virginia proved to be an appalling disaster. At first things must have improved for when the renowned Captain

Smith departed in late 1609/10 he stated that there were 200 settlers. What then happened has perhaps never been described in such horrific detail as in this Account of the period from November 1609/10 to May 1610/11, which became known as the 'starving time'.

No doubt for collective security the settlers had concentrated in James Fort but a siege imposed by Native Americans, no doubt in order to safeguard their own precious food stock from surprise raiding, ensured that they stayed there. Under blockade, James Fort must have developed all the attributes of a concentration camp with pitiable, listless, skeletal inmates trapped inside their palisade compound. The majority died of starvation over the next few months when hunger forced the inmates to turn to cannibalism by being forced to devour:

'those things which nature most abhors, the flesh and excrements of man as well as of our owne Nation'

'an Indian digged out of his grave after he had been buryed three days and wholly devoured'

'others eyeing the better state of body of any who hunger had not wasted so much as their own laid wait and threatened to kill and eat them'

one among the rest slew his wife as she slept in his bosom cut her in pieces, powdered her and fed upon her until he had devoured all parts except her head and was for so barbarous a fact and cruelly lustily executed'

Recent excavations at Jamestown have confirmed that cannibalism certainly took place.

In May 1610 of the *'Sea Venture'* castaways, who may well have included Richard and Isabel arrived in the *Patience'* and *'Deliverance'* built on Somers Island. This arrival of the survivors from Bermuda is described by the Ancient Planters though they suggest that they only numbered about 100, rather than 140 reported by Sir George Somers:

'the extremity of famine continued in the colony till the twentieth of May following (1610) when unexpected but happy arrival of Sir Thomas Gates and Sir George Somers in two smale barques which they had build on the

Somers ilands after the wrack of the Sea adventure wherein they sett forth from England with them one hundredth persons barely provided of victuals for themselves; they found the Colony consisting then of but sixty persons mostly famished and at the point of death of who many soon after dyed'

Almost immediately on arrival from Bermuda Acting Governor Gates decided that the best course of action would be to abandon Jamestown to the local Indians and make for Newfoundland. This order was only rescinded by the arrival on the 6th June 1610 from Cowes, England of the appointed Governor De La Warr with the 4rd Supply mission carrying 150 persons on the *'Blessing' 'Hercules'* and the *'De La Warr'.* His personal entourage included a personal guard of 50 Halberdiers, (one of whose halberd weapon, bearing the family crest of a Griffin has recently been excavated from a well at James Fort). The total number of colonists at this time would now have been just over 300.

Richard, as a skilled carpenter would have been much in demand to construct accommodation and to build defences, but he would also have sought his contracted allocation of 10 acres, probably on James Island, and built a cabin for the family as soon as possible. He would have been allowed an allocation of time – probably a day a week between May and September – to enable him to get his land planted up. Wages were laid down for the various skilled occupations. Colonists who were fed from the communal pool had a set deduction from their wages. Almost certainly Richard would have provided for the family. He had farming background and was an ideal settler for he could not only build and make but he could plant and grow.

Sir Thomas Gates departed for England almost immediately for reinforcements and supplies taking William Strachey's draft of the new disciplinary code. Strachey's subsequent report of the *'Sea Venture'* ship wreck and the amazing survival of those long since believed lost astounded London and perhaps inspired Shakespeare in *'The Tempest'.* Sir George Somers sailed back to Bermuda to collect food but he became unwell and died there on the 9th November 1610 aged 56. His heart is believed to be buried in Bermuda. In early 1611 De La Warr fell ill and departed leaving Capitaine George Piercie as a temporary leader in charge:

'a gentleman of honour and resoluton'

At the time of his departure in early 1611 De La Warr estimated:

The Ancient Planters' Witness Statement

'the number I left were about 200 most in health'.

This indicates the very short life expectancy in Virginia for it shows that perhaps 100 – a third of the colonists – had died in well under a year.

On May 19th 1611 Sir Thomas Dale arrived in Jamestown with 300 colonists, and as the most senior leader in Virginia he assumed the title of acting Governor pending the return of Sir Thomas Gates. Sir Thomas Gates returned as Lieutenant Governor on August 30th 1611 in the *'Trial' 'Sarah'* and the *'Swan'* with a further 300 new settlers. The size of the settlement would now have been around 750. It seems that there may have been a significant deterioration in quality, ability and attitude of a large number of the new colonists who arrived 1610/11, for Sir Thomas Dale reported that the 300 disorderly persons he took with him to Virginia were mutinous and unchristian and so disordered that only 60 of them were employable.

LORD DELAWARE.

That the initial short term perspective of the Virginia Company focused on seeking instant wealth is evidenced by De La Warr's immediate action on his arrival in June 1610. He ordered the closure of the two outlying forts and consolidated the colonists at Jamestown. Even though Jamestown was still suffering considerable shortages of food and stores he fitted out a 100 strong force to head for the mountains to prospect for gold and silver. It was only because the only mining engineers in the colony who were the key to finding suitable sites were killed by the Indians that the expedition was terminated. De La Warr became unwell and left within months of his arrival in March 1611, making it very clear in his letter of explanation to the Virginia Company in London that his honour should not be impugned for he was genuinely at death's door. The period from 1611 to 1616 under the Governorship of Sir Thomas Gates and his lieutenant Sir Thomas Dale was marked by the introduction of a brutal regime. In England William Strachey published the rules for military government under the title *'Lawes Divine, Morall and Martiall'.*

> *'Sir Thomas dale immediately upon his arrival to ad to ye extremity of miserie under which the colony had from its infancye groaned made and published most tirranous and cruell laws, exceeding the stricktest rules of marishall discipline'*

In the following five years the sanctions set out for indiscipline were ruthlessly applied. At Michaelmas, late September 1611, Sir Thomas Dale set off with 300 settlers 40 miles upriver. As a carpenter Richard would have been an important addition to this workforce but it is possible to speculate that it is more likely that he remained in Jamestown, for by now he would have established his house and holding with his wife and child. The centre of gravity of Richard's life in Virginia appears to have always been in the vicinity of Jamestown and if he managed to avoid conscription to the Henrico work gang this could have been due to the patronage of Sir Thomas Gates. If Richard was aboard the *'Sea Venture',* then he and Sir Thomas would have sweated together at the pumps to prevent the ship sinking, they may have established a bond from the shared experience of miraculous survival against all the odds and his exemplary contribution to building the two ships that enabled the castaways to get off Bermuda would surely have been recognised. Even though very much his social inferior, Richard may have developed a degree of personal chemistry with Sir

Thomas Gates and this may have saved him from being dispatched to Henrico and enabled him to avoid the worst of the excesses that went on there.

Employing the despotic rule authorised in the Virginia Company Manual, Sir Thomas Dale immediately set about construction for:

'the building of Henrico Town where being landed hee oppressed his whole Companie with such extraorinarie labours by day and watching by night as may seem incredible to ye eares"

The Henrico site was chosen as a replacement for Jamestown as the new main population centre because a bight in the course of the river makes it a virtual Island. It is a seven acre site connected by a neck of land to the northern mainland. Sir Thomas Dale built a strong palisade two miles inland, from river to river with a defensive trench two miles in length. However what is very apparent is that if the entrance can be breached then the site becomes a trap rather than a safe sanctuary.

The Report goes on to detail dreadful living conditions during the harsh winter with inadequate poor quality food:

'want of howses at first landing in the Cold of winter. And pinching hunger continuwally biting made those imposed labours most insufferable, and the best fruits & effects thereof to be noe better than the slaughter of his Majesties free subjects by starving, hanginge, burnyinge. Breaking upon the wheel and shooting'

It is significant that many of those subjected to this brutal treatment were the long term experienced artisans:

'havinge most of them already served the Colony six or seven yeare

The Report goes on to give specific examples of the brutality and inadequate rations endured by the settlers under Sir Thomas Dale:

'some for stealing to sattisfie the hunger weare hanged, and one chained to a tree till he starved to death'

'others attemting to runn away in a barge ... being discovered and prevented

weare shott to death, hanged and broken upone the wheele'

'besides continuall whipping extraordinary punishments, working as slaves in irons for terme of years (and that for petty offences) were daily executed many famished in holes and other poore Cabbins in the ground'

'Under this Tyrannous Government the Colony continued in extreme slavery and Miserie for the space of five yeares'

It is informative to contrast the condition of the 60 or so surviving colonists found in extreme misery by the survivors from Bermuda after the starving time in the spring of 1610, with the mental state of the settlers during this period of strict militaristic rule. In the first case the reports indicate that extreme hunger reduced the meagre population to pathetic zombie like, half human creatures suffering from extreme starvation and unable to think and act for themselves. Sir Thomas Gates judged them to be lazy and idle but it is much more likely that after enduring such conditions they were simply incapable of action or initiative. A number of them soon died even after help had arrived, and this legacy of a sustained period of insufficient sustenance probably demonstrates that Jamestown had acquired all the characteristics of a death camp. In contrast the colonists, who may possibly have included Richard and Isabel, who lived through the newly introduced harsher conditions set out in *'Lawes Divine Moral and Martiall',* were in effect in a work camp. Many were reportedly inadequately housed and fed but they were not actually starving, for it was surely in the interest of the Governor that the work force maintained the ability to perform.

The Ancient Planters' Declaration indicates that the colonists were reasonably docile and seemed to accept their conditions with resignation. The most dreaded fear was that of being sent to purgatory. Under the rule of Sir Thomas Dale many of the settlers may have come to believe that they had already arrived there. It is probable that they were cowed by failure or recognition of the sheer foolhardiness of any insurrection in the light of subsequent reprisals and harsh punishments. There is evidence that people may become de-sensitised after witnessing instances of frequent and continuous brutality. The extreme brutality meted out seemingly alike, to both artisan and labourer status settlers, as projects were driven on remorselessly to build an alternative settlement centre at Henrico and to develop the Bermuda Hundred into Charles City, must have been a cancerous growth in the midst of the colony. For those with the eyes to see there would have been continuous and episodic indications of a serious malaise. A society can tolerate only a certain amount of social damage before turning sour. The settlers would have had an in-built sense of fairness and justice and the martial regime must have been viewed by the Ancient Planters as grossly unfair. There would have been an acceptance of a reasonable tariff of punishment – let the punishment fit the crime – but sentences meted out were clearly excessively harsh.

Unlike human beings, seriously sick societies tend to have a lingering, rather than a finite death. This community must have festered with a sullen and disaffected work force without any sense of a common purpose. This would have been even worse than an outright revolt which perhaps might have been lanced like a boil, allowing things to move on more productively. Instead there would have been indifference, the worst of all worlds, for any initiative they did take to improve their situation would inevitably be crushed. For example this response to a petition to the Governor to be granted an allowance of time to produce more food themselves:

> *'At this time in all these labours the misery throughout the Colony, in the scarcity of food was equall with penurious and hard time of living, enforced and emboldened some to petition Sir Thomas Gates (then Governor) to graunt them that favour, that they might employ themselves in husbandrie, thereby that they and all others by plantinge of Corne might be fedd better then those supplied of voctuals which were sent from England would aforde to do so: which request of theirs was denied unless they would pay the yearely rent of three barrells of Corne and one month worke...'*

The response to this offer by the workforce to produce some food for themselves shows the intransigence of the regime, for Dale was not prepared to make any concession and insisted that any corn produced would be subjected to both a product tax and a labour tax. Richard would have been able to produce food for his family on the ten acres or so that he had been assigned as part of his employment contract. Even if he had managed to cultivate this significant area of land it must have been too large for him to have farmed on his own so perhaps he allowed the landless labourers to work some of it in their meagre free time in exchange for some help. The scarce factor of production in Virginia was likely to have been labour rather than land. Life in Henrico City under the rule of Sir

Thomas Dale was nasty, brutish and frequently short. The settlers seem to have become acquiescent to their situation and in spite of their degradation and pitiful state they apparently retained hope for the future and certainly retained the mental capacity to seize the opportunity when it came. Immediately before Sir Thomas Gates left Virginia in 1614 on the *'Elizabeth',* at the instigation of Sir Thomas Dale, the Ancient Planters report the momentous potential change in their fortunes when it was agreed that:

> *Upon the promise of absolute freedom after three more years to be expired (having most of them already served the Colony for six or seven years in that general slavery) were yet contented to work on the building of Charles City and hundred with very little allowance of Clothing or victuall'*

Though this promise had no legitimacy or substance whatsoever and was agreed under coercive conditions, it was an offer the Ancient Planters were in no position to refuse. Because of their circumstances in all likelihood they would have felt compelled to acquiesce to the Company's demands. By 1615/16 many of the Ancient Planters would have completed their contracted time service to the Company and the offer only extended this period without recompense. The proposal amounted to an illegal extension of the individual's contract with the Virginia Company without fair choice. The reward for agreement was the promise, which given the previous track record of the Company was of dubious reliability, that the Ancient Planters would be granted their originally contracted condition of release from Contract three years late. But the Ancient Planters were powerless and in no position to reject the proposal. The reason that these conditions were accepted suggests that the colonists after being subjected to the harsh regime for so many years must have been cowed into submission and desperate to survive, for at first sight it is otherwise difficult to understand why they accepted Sir Thomas Dale's proposal.

But perhaps it was in the circumstances actually a rational choice to go along with the agreement. These were among the original settlers who had invested the best years of their lives to the Colony. They had built their houses with their own hands, developed their small productive plots in any free time allowed and established their family home. They comprised the skilled, permanent core workforce but were unable to exert the potential influence of this position from the fact that a greater need was survival of themselves and their family. If they did not accept the proposal they knew that they were vulnerable to repercussions from the Virginia Company. They had nowhere else to go and without the protection and supplies from the Company they could well perish. These are the most obvious reasons why the conditions were probably accepted willingly and even with alacrity. But outweighing all of these was that the proposal included the key ingredient, the tantalizing prize of hope for a better future.

The colony may have been able to struggle on for years with a static or falling population eking out a subsistence living. But that would have been no use to the Virginia Company. Behind all the rhetoric and fine stated aspirations this was a mercenary institution that had to satisfy its increasingly restive investors who were more interested in getting their money out than putting any more capital up. The tipping point came in 1616. Something had to change – and it did. There was some good news during the period of harsh rule between 1611-1616, for it marked the end of the First Powhatan war which was followed by the development a more harmonious relationship with the local Indians, particularly after the marriage in 1614 of John Rolfe and Pocahontas, daughter of the Powhatan tribal leader. A line may therefore be drawn in the sand with the departure for England in 1616 of Sir Thomas Dale accompanied by John Rolfe, his wife Pocahontas and their year old son Thomas. This point signals a sea change in organisational philosophy with a departure from the militaristic centralised style rule to a decentralised laissez faire approach. The Declaration then records the Virginia Company history after this change in the philosophy of the Virginia Company and describes the consequences of how it was managed. It includes an account of the massive increase in the number of colonists arriving immediately before the massacre of March 22nd 1621/22.

In 1616 the Ancient Planters, including both Richard and Isabel, were offered what proved to be an extremely powerful and attractive incentive. The deal was that if the Ancient Planters agreed to continue to work for the Virginia Company for a further three years they would not only be freed from any further obligation to work for the Virginia Company but they would then be given a dividend of land for their 'Personal Adventure'. This proposal was only formally promulgated in 1619 so during the immediately preceding three year period the Ancient Planters must have been in a frenzy of apprehension as to whether the promise would ever actually come to fruition but their fears proved groundless for the Company kept their word. In April 1619 Sir George Yeardley issued a Proclamation from the Virginia Company which freed all residents in Virginia before the departure of Sir

Thomas Dale from bondage to the Company and granted 100 acres to each Ancient Planter in recognition of their contribution of service. A second major initiative was the introduction of the head right system. Those who arrived after the departure of Sir Thomas Dale were entitled to a lesser grant of 50 acres and so landowners who paid the transportation costs of a labourer were entitled to the 50 acres, not the indentured servant. This was a New Deal indeed. There must have been growing anticipation, hope and excitement among the Ancient Planters prior to the Proclamation. The effect was immediate with:

> *'all of them following their particular labours with singular alacrity'*

> *'free liberty was given to all men to make their choice of their Dividends of lande'*

> *'within the space of three years our Countrye flourished with many new erected Plantations from the head of the River to Kicoughtan beautiful and pleasant to the spectator and comfortable for the relief and succour of all such as by occacion did travaile by land or watter everyman giving free entertainment to friends and others'*

The Declaration describes a bountiful and almost idyllic situation with plentiful, corn, cattle, swine, poultry and provisions. Vines and mulberry trees were planted and trials set up for new varieties of English corn and silk grass. But the Report tends to gloss over two key matters that give a less flattering perspective to the behaviour of the Ancient Planters. Firstly, bearing in mind that they were themselves by now probably fairly immune from the common diseases in Virginia, there is only a passing reference to the awful mortality among the new arrivals who:

> *'arrived at the most unseasonable time of the yeare beinge att the heat of the Summer and divers of the shipps brought with them the most pestilent infections whereof many of their people had dyed at Sea so that these times of plenty and liberty were mixed with the Calamities of sickness and Mortality'*

The second major issue is that of the greed and self interest displayed in choosing and securing their land dividend which, by provoking the massacre, proved to be a near fatal cancer at the heart of the Colony. This is dismissed in a one sentence comment:

'justly likewise we were punished for our greedy desires of presente gaine and profit wherein many showed themselves insatiable and Covetous'

There is an implicit recognition that the greedy seizure of land without agreement with the indigenous population, but without their overt protest, may have been interpreted as a fatalistic acceptance. In a letter dated January 1619/20 John Rolfe reflects this complacency:

'All the Ancient Planters being sett free have chosen places for their dividends according to the Comysion. Which gives all greate content for now knowing their owne lands they strive and are prepared to build houses & to clear their groundes ready to plant, which gives great encouragement and the greatest hope to make the Colony florrish that ever yet happened to them'

The settlers had clearly lured themselves into believing that the bad times were now past but there was now to be the rudest of awakenings:

'we being too secure in trusting our treacherous enemy the Savadgas, they whilest we entertained them friendly in theire houses tooke theire opportunities, and suddenly fell upon us, killing and Murtheringe very many of our people, burning and destroyinge theire houses and Plantations'

The devastating effect that the attack had on the colony is shown in the 'cri de coeur' statement written about the effect of the attack that it:

'strucke so att the life of our welfare by blood and spoile, that it almost generally defaced the beauty of the whole Colony puttinge us out of the way of bringeinge to perfection those excellent works wherein we had made so faire a beginninge'

CHAPTER FOUR

The Virginia Company & The Valley of Death

4

FOR MOST OF its short life the Virginia Company walked through the Valley of Death. In modern management parlance this term is used to describe a catch 22 situation when a small Company, without a viable track record of proven success, is unable to raise sufficient additional finance to support what it perceives as a golden opportunity. This comes about because the once bitten twice shy previous subscribers are hesitant to share the Company's optimism and are consequently reluctant to put up more risk capital. In the end the Virginia Company was able to acquire surrogate capital by distributions from its land bank and emerged from this perilous journey through the Valley, though it was touch and go at times. But having managed to survive, its progress was then punctuated by the massacre of March 1621/22, which, but for a timely warning from a Native American passed to James Fort by Richard Pace, would have wiped out the Colony. The great majority of the settlers who had subscribed their life chances were not so fortunate for they died in the first permanent settlement to be established in the New World. During the seventeen years between 1607- 1624 perhaps 10,000 colonists shipped to Jamestown. The attrition rate was awful with a death rate of around 85%, the great majority dying at most within a year or two of arrival.

This chapter weighs the evidence presented in the Ancient Planters' Witness Statement and comes to a judgement regarding the actions of the Virginia Company as they impacted on the lives of Richard and Isabel. The Ancient Planters' Declaration described what happened. This is an attempt to understand why it happened. In addition, the remarkable amount of information now available about the activities and development over time of the Virginia Company is used to complement the assessment – if anything there is a problem of too much data. This was a functioning bureaucracy and although the quality of decision making may have been abysmal the details registered and the maintenance of the accounts was diligent. Sufficient records have survived to subject them to examination under a critical lens in order to gain an insight into what actually occurred and the probable reasons for the events.

Since the first settlers landed in 1607 the Virginia Company had an unspectacular beginning. In 1616 it was still in a Valley of Death situation. By this time the Virginia Company's investors were disillusioned and its reputation severely compromised. Rumour of the troubles in the colony must have leaked back to London and this would have deterred any ready supply of willing worker recruits. In desperation the Virginia Company took decisions that, though they probably didn't realise it at the time, were to have enormous consequences for the survival of the settlement:

1. Sir Thomas Dale returned home in April 1616 and George Yeardley took over as Acting Governor. John Rolfe sailed back to England with Sir Thomas accompanied by his wife Pocahontas and their infant son. This visit was clearly intended to promote the virtues of both the Virginia Company and the potentially profitable opportunity offered by the developing tobacco trade.
2. Unable to pay the due dividend in cash the Virginia Company announced a dividend of 100 acres for every share held. Ancient Planters were each to be regarded as owning one share. The share with the attached land entitlement could be bought and sold.
3. The head right system granted 50 acres to anyone who transported an indentured servant or worker to Virginia at their personal expense.

By far the most significant outcome of these changes resulted from granting private ownership of land to Ancient Planters, now freed from all ties and obligations to the Virginia Company, and also to any newly arriving settlers who had acquired territory from the land dividend. After long years of servitude this opportunity set alight a flame of entrepreneurial ownership that swept all before it. Over 80 plantations were established or expanded up and down both banks of the James River. During the course of the initial development frenzy many of the new arrivals died and eventually the Native Americans lost most of their tribal homeland. Without doubt this shift in power was the trigger for the transformation in the fortunes of the colony and the key game changer. Eventually the white settlement of Virginia was reborn on a stable and prosperous future. Richard and Isabel lived through these momentous events.

The final third of Richard's life from 1609/10 to 1622/23 was inextricably linked to the Virginia Company. Through this relationship he would experience remarkable events and witness terrible scenes, he would have his hopes raised and then cruelly dashed, yet finally he would have the chance to achieve significant prosperity. Richard is probably best described as an indentured Virginia Company Adventurer. There were two broad categories of Adventurer. Investor Adventurers were those who put up the capital through subscribing to an issue of shares, to finance expeditions to Virginia. The other main group, which included some of the first category, were those who actually voyaged to America. The first group risked their money while the second group risked their lives. Power and control of the strategic decision making lay predominantly with the 'stay at home' Adventurers, who, by definition would have had little practical experience of what a pioneer life entailed. The consequence of this lack of knowledge was suffered by the settlers in Virginia.

The term 'Adventurers' is certainly an appropriate way to describe the first settlers who arrived in Virginia beginning in May 1607 for it implies engaging in a daring, hazardous and speculative activity. The chance of success would have been enhanced by a leadership which was able to promote a vision of purpose, which had good organisation and planning and the ability to raise sufficient finance to ensure resources were adequate for the enterprise to at least have a chance of success. Unfortunately the Virginia Company lacked these qualities. The passenger lists for the first three voyages to Virginia, before Richard and Isabel were involved, clearly shows that the Company failed to recruit people with appropriate skills and abilities required to achieve the task before them. The early settlers were a mixed bag of inexperienced people who were, poorly equipped, and, with the exception of Captain John Smith, lacked leadership ability.

The subscription list for the Adventurer group who bought shares in the Virginia Company was composed of members of the aristocracy and also included many of the London Worshipful Guilds with other contributions drawn mainly from prosperous gentlemen and well off tradesmen The Board of the Company was run by an aristocratic and elite group predominantly occupying that position by brute luck, such as accident of birth, wealth or sinecure. A major problem was the huge gap of understanding between the Virginia Company based in London and the lived experience of those 3000 miles away in Virginia. At the least, particularly with such a long communication lag, this would have required adaptability in decision making and a capability to match information on a changing situation with appropriate action. As information filtered back to London it seems that, rather than spearheading new thinking, no reappraisal was made. Certainly there was an insufficient response because the Board, set in their way of looking at the world had little understanding of the conditions in Virginia and seem to have been overly optimistic. When they did eventually yield to the need for change and greater support, the actions that they initiated were often ill considered. Any attempt at settlement was always going to be a high risk venture with probably no more than an evenly balanced chance of success. The Board, rooted in the past with a frame of reference conditioned by an aristocratic mentality and expected customary relationships where obedience to instruction could be enforced by harsh sanctions, lengthened the odds. There were three broad approaches to settlement over the seventeen year life of the Virginia Company 1607 – 1624:

1. The FINDING model which ran from 1607–1610 and ended with De La Warr's 4th Supply expedition failing to find any gold.
2. The BUILDING model symbolised by Sir Thomas Dale which ran initially from 1611 to 1616 but was then extended to 1618/19. This aimed to develop a defendable position by converting the toehold colony situated at Jamestown into a stronghold settlement at Henrico City and Bermuda Hundred.
3. The PRODUCTION model ran from 1619 to 1624 introduced private enterprise and private property rights in order to allow the exploitation of an innovative tobacco growing industry.

During the FINDING phase the Virginia Company at first also appears to have shown an aggressively self confident swagger in the manner in which they undertook the whole enterprise, taking wild risks and short cuts which they hoped would yield them quick money. An enterprise run by brute luck incompetents may survive for decades if the underlying economic source is abundant, but in the early days of Virginia there was no more than a potential for profit. An Adventurer having paid over his capital was given a contract document certificate guaranteeing that he would be given his dividend of any treasure or other wealth that might be found:

Shall have – according to his adventure his full parte of all such lands, tenemens hereditamens as shalle from tyme to tyme be ther recovered,

> *plaunted and inhabited. And of all such Mynes,& Mynneralls of golde silver and other Mettalls or treasure, Pearles, pretious stones or any kinde of whatsoever, merchandizes, comodities or profits whatsoever which shall be obtained or gotten in the said Voyage*

It is apparent from the Contract that the subscribers were not interested in a long term investment. The longer term is associated with greater uncertainty and there would have been far less enthusiasm if the subscribers believed that this would be the likely outcome. They were locked into the investment for there was no stock market to enable them to instantly liquefy their outlay. The model in their minds must have been the Spanish conquistadors and the expectation of matching the rich pickings that had been extracted from South America. But both the subscribers who put up the money and those Adventurers who signed up for this enticing package and then sailed for Virginia were to be sadly disappointed, for there was no instant jackpot. When it became apparent that short term returns were not forthcoming the investors grew increasingly edgy and dispirited and not at all inclined to throw good money after bad. The high status combination of royalty, aristocracy, church, livery companies prosperous merchants and traders who had invested made up a powerful lobby. They would have been keen to enhance their status by being seen to be involved in an extravagant enterprise, but not at all keen to be associated with anything with the taint of failure. Their increasingly hostile attitude was criticized by Lord De La Warr after his return:

> *'I perceive that since my coming into England such a coldnesse and irresolution is bred in many of the Adventurers that some of them seeke to withdraw their payments'*

When no treasure trove was found the subscribers became restless. Some samples of ore which were eventually sent back to England proved to be worthless. So in the early years the perspective was essentially short term with the Virginia Company focused on myopically chasing El Dorado in the hope of enriching the investors beyond their wildest dreams. During this period the colony was insecure and in constant danger of being wiped out by sickness, starvation or murderous Native Americans, and survived only by the skin of its teeth. The first three voyages transported something over 200 people under a squabbling leadership. By the winter of 1609/10 apart from a handful in

Jamestown itself, the colony was reduced to 60 survivors living mainly either among the indigenous population or dispersed to scavenge the oyster beds.

The new Letters Patent granted to the Virginia Company in 1609 offered the chance for a new beginning and it allowed them monopoly rights to take the lead in the creation of a new society in Virginia. This signalled the transition to the BUILDING phase of development. Over the next few years until at least 1616 the attempt to carry out this task must be judged an abject failure. Leadership in any such bold enterprise necessarily required a strategic plan, an effective organisational structure, task competence and the ability to develop a motivated work force. Instead the Company proved to be demonstrably incompetent. There was no dissemination of a wider sense of social purpose apart from service to the Almighty. Even given seventeenth century values, there was no indication of a basic humanity in the stewardship and concern for the artisans and labourers they had recruited.

Richard and Isabel arrived probably only shortly before the time of the imposition of strict militaristic rule. By 1611/12 the perspective of the Virginia Company had shifted away from immediate enrichment and proceeded with the development of the infrastructure with a new more suitable population centre first at Henrico City, and later at Bermuda Hundred. It was proposed that a College should be established in the vicinity for the joint education of both Settler and Indian children. The development took place over the next five years of despotic rule. Infrastructure development and the educational initiative may have been worthy causes but the fatal flaw was that they did not generate profits to satisfy the shareholders.

The England from which the settlers had formerly belonged was a very unequal society and this was similarly reflected in the composition of the Virginia Colony. The feudal system typically involved accepted mutual rights and obligations between the Lord of the Manor and villagers which were exercised through customary rights, copyhold tenancies and the Manor Court. A Lord of the Manor had the right to conscript tenants into his force to support the King, he was entitled to acquire the best beast when the Copyhold was transferred on death, and his Steward through the Manor Court could determine whether or not a significant fine should be paid for misdemeanours such as not maintaining gates or throwing ditches. This format would have remained in the mind-set of the settlers now living in Virginia and they must have perceived the absence of the opportunity or machinery for a redress of grievances. During the years 1611-1616, Sir Thomas Dale did not engage with his workforce and so his rule became one of destructive inequality. Colonisation implies a situation when an occupying power exploits the local population. Dale exploited his own men and drove them in a manner that amounted to internal colonisation. There are obvious signs that during the period this must have been a dysfunctional society with a lack of trust, respect or sense of purpose between leadership and workforce. Even with the known harsh punishments, some of the workforce still tried to steal food, attempted to escape in small craft to the wild ocean, chose to abscond or even preferred to live with the indigenous tribes.

By 1616 the Virginia Company was in deep trouble and facing a crisis. It was making no money and it had a demoralised and insufficient workforce to exploit the potential of the colony which was only now becoming increasingly apparent. In June 1616 the colony was still tiny, with only 365 settlers. Until now the near Imperial standing of the Virginia Company had allowed it to operate with few restraints on its actions as long as it retained the confidence of its investors. This status had so far allowed it to exercise unaccountable arbitrary power and to operate as a pseudo-monarchical fiefdom. Up to now it had also demonstrated a remarkable ability to survive its calamitous bungling. In reality the Virginia Company had metamorphosed into a monster of incompetent organisational autocracy which treated its own employees almost as slaves except the Company would have continued to pay wages. By now it must have been glaringly obvious the Company was on the road to nowhere and required re-invention rather than a makeover. Re-invention meant that it had to generate profit. The absence of a viable colony after ten hard years was a reflection of how the Virginia Company operated, how it was owned, how it was financed and incentivised within a societal framework. The Company needed to encourage innovation and to raise finance to cover the investment to exploit its potential. It also required an ownership structure that could capitalise on any opportunities and enlarge the scope for innovation and investment and which could demonstrate, particularly after so much suffering had been endured, that it was not blind to its social obligations. The track record suggests that it is hardly credible that the Company sat down and considered the problem and set out a rational plan in order to overcome the crisis. The volte face in allowing the radical PRODUCTION based approach has all the characteristics of a reluctant decision of last resort. But there was an invisible hand – in fact two invisible hands – that did come to the rescue. These may be labelled 'Serendipity' which conquered the innovation issue and

'Desperation' which dealt with the problem of lack of finance and forced a reform of ownership. Unfortunately a consideration of social obligations was not adequately addressed until after many of the settlers had perished.

Innovation

The innovative force is indelibly associated with the name John Rolfe. Though prosperity came to be somewhat dangerously based on the single staple tobacco crop there is no doubt that this was the only candidate around able to have provided the necessary innovative stimulus. Gathering all the threads of what took place together gives an understanding of the underlying process at work and allows an insight to the extraordinary and fortuitous serendipity that allowed the development of a viable tobacco industry that almost certainly saved the Virginia Settlement. Facilitating this particular innovation required three steps:

> The Science – developing an appropriate strain of tobacco seed suitable to local conditions through selection and trials.
>
> The Technology – planting, growing, cropping, drying, packing and shipping of the tobacco leaf.
>
> Ownership – a structure able to provide the entrepreneurial drive needed to exploit the opportunity and able to raise the finance to supply the necessary resources in a supportive environment.

Either by vision or happy accident Rolfe had experimented with tobacco which had previously been controlled on European markets by the Spanish. He had acquired from the Caribbean a particular strain of tobacco seed that thrived in the climate of Virginia and which became better liked by European smokers than 'Spanish' leaf. However in order to translate an innovative capability to commercial production needed the requisite technology. From around 1612 Rolfe and others had grown and experimented with tobacco crops on a small but increasing scale between 1612 /1616 under the Sir Thomas Dale regime. The Secretary of Virginia later wrote:

> *'I may not forget the gentleman worthie of much commendations which he first took the pains to make trial thereof, his name John Rolfe, Anno Domini 1612, partly for the love he hath a long time borne unto it and partly to raise commodity to the adventurers'*

It was not until 1616 that there is any documentary reference to a commercial tobacco crop. In these intervening years the technology of growing and treating the leaf would have been developed. Rolfe gave some tobacco from his crop to friends:

'to make trial of ... and it ... smoked pleasant sweet and strong'

Richard and Isabel may have been involved, perhaps growing trial crops on their own small acreage. Now that an innovative product winner was on the verge of commercial production, in order to break out of the valley of death the Virginia Company had to raise sufficient finance to recruit a large enough workforce to produce it. To do this the Company had to now promote itself, and extol its profitable potential to investors. The realisation that this was required was signalled when John Rolfe sailed from Jamestown in 1616 on board the *'Treasurer'*, accompanied by his wife and Thomas their infant son. After much anguished soul searching Rolfe had married Pocahontas on 15th April 1614. Richard and Isabel would almost certainly have been at their wedding. Rolfe's marriage and the birth of their son Thomas to a mixed parentage seems to have led the colonists to believe that perhaps a peaceful coexistence with the Indians was possible After the union her father rescinded the instruction to the Confederacy Tribes to kill any white man on sight and this ended the 1st Powhatan War. There is no doubt that John Rolfe had genuine feelings for her:

'It is Pocahontas to whom my hearty and best thoughts are, and have been for a long time so entangled and enthralled in so intricate a labyrinth that I could not unwind myself thereout'

The presentation of an Indian Princess to an incredulous English society symbolised the change in the Virginia Company strategy. The visit was very unlikely to have been just for personal reasons but was rather a premeditated marketing initiative calculated to re-brand the Virginia Company in order to promote the opportunities of colonisation and raise some cash to stave off ruin. Pocahontas was the jewel in the Crown. One can imagine London gossip recounting that she was the favourite daughter of the Powhatan tribal chief who was believed to have fathered over a hundred children. In England the story of the marriage of a Native American teenager to the educated and articulate John Rolfe and her visit to England where she was treated as royalty would have been surreal. This young wife who was unlikely to have ever travelled more than handful of miles from the James River had now travelled 3000 miles across the Atlantic The story would have spread like wildfire of how she had been abducted in 1612 and held hostage in Jamestown by the English. How this wild running girl had been renamed Rebecca, and adopted western dress. How she had been taught the English language and converted to Christianity. And finally, adorned with the trappings of royalty, how she had been presented to King James I and the Queen at Hampton Court Palace. The visit could be sold to English society as a theatrical fairy tale which demonstrated living proof of the success of the stated objectives of the Virginia Company. It is very probable that on his return to England while he was living at Brentford, Middlesex, only just over a mile down river from Kingston-on-Thames, John Rolfe would have communicated with Richard's younger brother Thomas and let him know how he was faring in Virginia. It is also likely that he took his new wife Pocahontas and son, to be welcomed by his old friends and attended a thanksgiving service for his safe return at St James Clerkenwell. It was surely no coincidence that it was at St James, Clerkenwell, and a long way from Jamestown, that their son Thomas Rolfe would marry Elizabeth Washington in September 1632 But the marketing initiative was to have a tragic ending. Towards the end of May 1617 the Rolfe family set off from London to return to Virginia but as they sailed down the Thames a sick Pocahontas had to be taken ashore at Gravesend where she died, aged just 22. She was buried there at St George's Church. The Parish Registry entry shows her husband as Thomas not John. There is a modern memorial statue in the churchyard. No matter how the visit seemed contrived to support the commercial ambitions of the Virginia Company there can be little doubt that John Rolfe was clearly very distressed about her death:

'great is my loss and much my sorrow to be deprived of so greate a comfort ... in her whose soule (I doubt not), resteth in eternal happyness'

John sailed on with his son and on April 10th after a smooth passage they arrived in Plymouth. Thomas was very sickly and was landed there in the charge of John Rolfe's brother while he returned to Jamestown. He had lost his wife and now had to leave his only son who he would never see again. It was probably her vulnerability to new strains of infection that had done for Pocahontas. Rolfe was particularly concerned that his action might bring censure from his peers and so he wrote a detailed account to justify his actions. The ship then made fast passage to Jamestown arriving just over a month later to end an eventful and fateful visit. But overall the Company must have chalked up the marketing initiative as a great success. Virginia was back on the map.

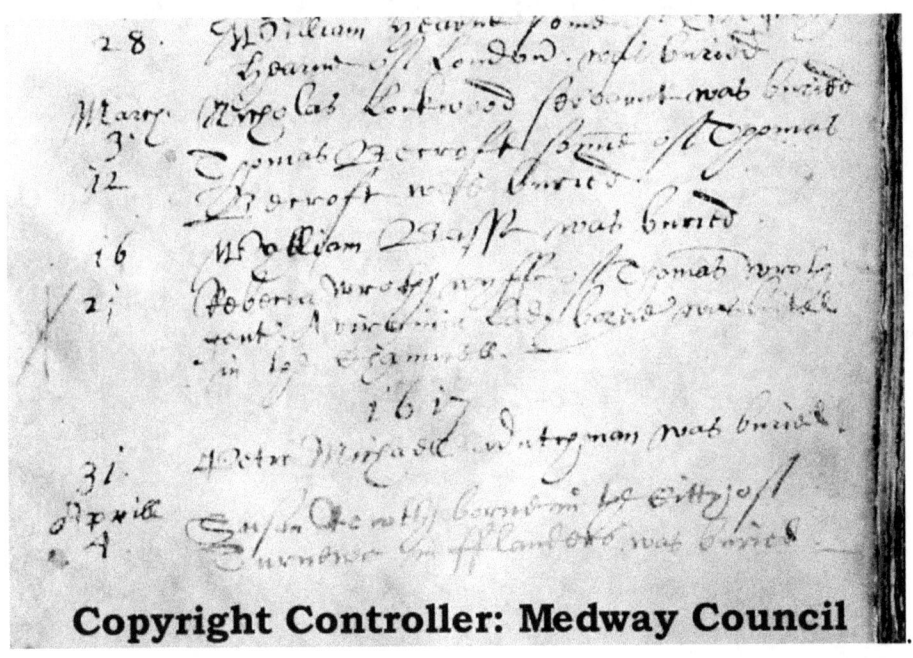

21st March 1616/17.
Rebecca Wroth wyffe of Thomas Wroth gent A virginian Lady borne was buried in the chauncell

POCAHONTAS

This engraving is the only known portrait of Pocahontas rendered from life. During her stay in England, Dutch engraver Simon van de Passe captured her likeness and recorded that she, like the artist himself, was 21 years old. It was the first of many depictions of Pocahontas intended to demonstrate that a Native American could adopt the demeanor of a "civilized" European. The Virginia Company – backers of the Jamestown settlement – likely commissioned the engraving with this in mind, hoping to attract more colonists and investors. The image also promotes the false impression that she was a princess in the European sense; the inscription describes her as the daughter of a mighty emperor, and the ostrich feather in her hand is a symbol of royalty. But this engraving offers a sound estimate of Pocahontas's true appearance.

The Virginia Co. & The Valley of Death

On his return Rolfe reported his impression of a happy settlement with the colonists poor but relatively content, in sharp contrast to the previous harsh regime under Sir Thomas Dale. Rolfe, now appointed as the Company Secretary, wrote to Sir Edwyn Sandys in London on June 8 1617:

'All men cheerefully labor about their grounds, their harts and hands not ceasing from worke, though many have scarce rages to cover their naked bodyes. English wheate, batly, Indyan Corne, Tobacco greate plenty in the ground. Hemp and flax seed distributed by the Governor and is putt in the grounde: nothing neglected which in any waies may be available to advance the Colony'

John Rolfe also reported Chief Powhatan's fatalistic acceptance of his favourite daughter's death and again emphasised the happy state of affairs he found on his return to Jamestown:

'he laments his daughter's death but glad her child is healthy and wants to see him but desires that he may be stronger before he returns'

'the great blessings of God have followed this peace and it next under him hath bredde our plenty – every man sitting under his fig tree safely gathering and reaping the fruits of their labours with much joy and comfort'

This comment is revealing for although it suggests reasonably contented settlers they do not appear to be a dynamic group anxious to seize opportunity. Secondly it is most significant that it still only consisted of a total of 365 people of whom the 150 or so Ancient Planters would have been the vital skilled core. Even of the group had exhibited some dynamism there were far too few of them to have produced tobacco on a commercial scale. Around this time John Rolfe also described the structure of the colony as strictly demarcated on class lines. About a third were gentlemen, supported by the military who were required to work only one day a month for the communal good A further third was made up of skilled artisans and a third of labourers. An essential first requirement for commercial tobacco production was a huge increase in the working population. To recruit and ship them out would require a lot of finance from cynical and disenchanted investors who were now expecting payment of their first dividend on the due date – a payment that the Virginia

Company lacked the cash to make. However the Company had one asset in abundance with which they were able to pay a surrogate dividend – and that was a vast land bank.

Finance

Innovation involves risk and needs paying for. Exploitation of the labour intensive tobacco industry required possibly a ten-fold increase in workers. The initiative in 1618/19 to introduce a head right system became the primary means by which the necessary workforce was recruited. The only currency at the Company's disposal to finance the growth required was its vast land bank of Virginia, so this was used as a surrogate form of finance. The announcement of the scheme in a short rather innocuous sounding statement would have a profound effect on the colony and the lives of thousands of emigrants.

> '50 acres for every person which at their own charges they have transported to inhabit Virginia before 24 day of June 1625 if he continues there for 3 years either at one or several times or dies after he is shipped for that voyage'

Landowners who arrived after April 1616 and paid their own passage received 50 acres for themselves and a further 50 acres for every person they transported at their own cost. Richard and Isabel themselves would choose to exploit this provision in 1621 when they returned to England to recruit six workers to work their plantation for which they eventually received a grant of 300 acres The initiative was a huge success as far as emigrant numbers were concerned but for the great majority of the new arrivals it would prove to be a disaster.

In 1617 the total population was around 350. In 1619, 1620, 1621 a total of 3520 arrived in 42 ships which employed 1200 seamen. In 1622 a further 1000 arrived. The shortcomings of the London based Virginia Company's scheme are obvious and would have grave consequences for incoming settlers who were ill prepared for the harsh conditions. For instance waifs and young women were sent out who were totally unfitted for a pioneer life. It is said a revolution eats its children. One source of new emigrants was the Bridewell, an institution that had been set up under Elizabeth I for the education of destitute children, the care of paupers and the occupation of vagrants. It had later become a correctional institution from which apprentices could be cheaply bound and so it became a fruitful recruitment source. In the same vein the Mayor of London scoured London for stray children and shipped them to Virginia so alleviating any charge on the Parish under the Elizabethan Poor Law. The contribution to the population of Virginia from the London prisons and penal institutions was also considerable. Convicts and about to be convicted felons were also shipped out. In 1621/22 a hundred young women from respectable homes who would have had little idea of the privations of life in Virginia were kitted out by a group of Adventurers and had their passage paid to become brides. Tobacco appears to have been the surrogate currency and so the fee to the prospective husband was a charge of 120lbs of tobacco. This charge was raised to 150lbs for the next batch of young women as it was claimed that the quality of the leaf had deteriorated during shipment.

The mortality figures for 1619/22 are truly terrible. The calculations appear credible and require to be quoted in detail in order to give a picture of the death rate. The Records of the Virginia Company show that:

In the three last years of 1619, 1620, and 1621, there hath beene provided and sent to Virginia 42 sale ships, 3570 men and women for Plantation and in those ships above 1200 employed

In 1622 there is a note by Mr Wrott, endorsed by Sir Nathaniel Rich, detailing for each of the three years the numbers and the likely the fate of these people:

'the first list 1619 agrees with my first calculation to a man – 887 persons'

'The second list for the yeare 1620 taken at the latter part of the yeare or the beginning of 1621 amounted to 843 persons whereof about 240 have their names Crossed ... it was confessed that they were all dead so the remainder is few more than 600 and then of this list it appears that above 120 persons ran away or dyed in their passage'

'The third list 1621 was by the first calculation 1472 but on second reviewe we find it to bee 1501'

'The 4th list taken 1622 about the time of the Massacre we find about 1240'

'In the years 1619; 1620; 1621; there was 3560 or 3570 Persons transported to Virginia and Sir Thomas Smith left above 700 Persons which makes 4270 Persons whereof the Remainder being 1240 about the tyme of the Massacre it consequentlie followes that we lost 3000 persons within those 3 yeares'

'And that in the latter end of the yeare 1622 there were transported neare upon 1000 Persons whereof manie dyed by the way and it appeared that by the sword and sickness there are perished 500 since the massacre. So that by this account there cannot be more than 1700 Persons in the Colonie'

This graph is based as far as possible on figures found in the documentation though some estimation has been necessary to produce a continuous series. It is apparent that, apart from the possibility of escape offered by the nine vessel 3rd Supply fleet in June 1609, the Virginia Company had been trapped in a valley of death phase until 1618/19. The graph confirms how very small the settlement was from its inception until 1617/18. Though a high proportion of the resident population died during the starving time of the winter of 1609/10, they accounted for a relatively small number of the total deaths during the Virginia Company's existence. There is a significant increase in the population 1611 with the arrival of the 600 brought in by Dale and Gates but then the colony gradually declines under militaristic rule with few new emigrants. In 1617/18 it is only a little larger than in the early years. The graph shows the dramatic rise in population in 1618/19 as a result of the introduction of the head right system and the associated large increase in the death rate. Though eventually the Virginia Company succeeded in passing through the valley of death phase the passage involved a massive increase in deaths. But the success was achieved only because the supply of new labour exceeded the death rate.

The mortality figures suggest it is necessary to be wary of some of the ambience that exudes from the Ancient Planters' Declaration. Though it does refer to the fact that the *'times of plenty were mixed with the Calamities of sickness and Mortality'* it did not spell out the extent of the mortality. There is an obvious and huge contradiction between the picture of the Ancient Planters socialising with their peers up and down the James River while the bonded employees were dying in their hundreds. Most of the Ancient Planters would by now have been immune from many of the maladies the newer arrivals contracted, so the high mortality may have been less of a personal concern. Because the Declaration was produced by a sectional interest it may not have revealed the full story. The Ancient Planters were the new elite, enjoying equivalent land rights to the old money and they would have been in a much happier situation and frame of mind than those shipped out under a false prospectus without being informed of the risk to life or pressed into transportation against their will. It is possible to speculate – and it can be no more than a hunch – that

Richard and Isabel together with some similarly devout old friends, separated themselves in a grouping in the vicinity of Pace's-Paine in order to escape the excesses of this surge of self interest and to cohabit with persons of a similar moral standing as themselves.

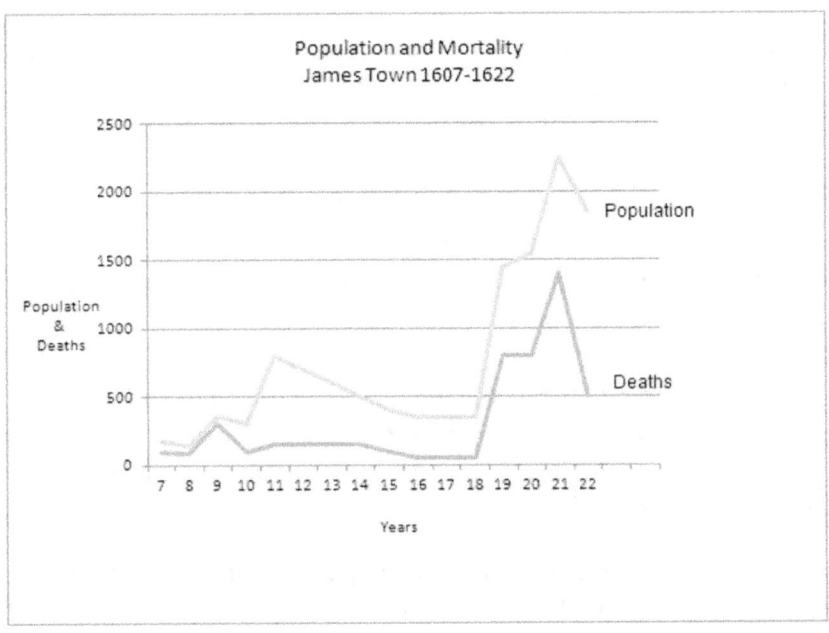

The influx would have provided a workforce for the tobacco industry although most did not live for more than a year or so. The mortality figures are evidence of the less fortunate workforce, who would have included many of the young women shipped in for marriage, who toiled and lived perhaps for a year or two before they succumbed to the hostile environment and were then replaced by the next shipload. Richard and Isabel were Virginian old hands and it is no coincidence that when they came back he selected and recruited their own workforce who they judged would be able to survive and perform in the colony. Richard ensured that by paying her fare himself, his niece Ursula Clawson would not be subjected to the marriage auction. The common infectious diseases were plague, tuberculosis, measles and small pox. When sickness broke out on outward bound ships it seems there were no effective quarantine arrangements put in place on reaching Virginia. Certainly such precautions were known about for in 1596 the City of London had been successfully protected from a plague epidemic sweeping the rest of London. Given the callous attitudes of the times where life seems to have been cheap, when the Board of the Virginia Company later calculated the figures they suggested that they must:

'with much humility acknowledge the just finger of Almighty God...and by future amendment, in better attending Divine worship and more caerfully observing his holy and just Lawes'

Arrival arrangements were haphazard with an insufficient provision of reception accommodation in spite of pious entreaties from London, though noticeably not backed by an offer of the necessary finance, to construct such accommodation. Plantations were instructed:

'shall each of them at their common charge, labour and industry, frame, build, and perfect with all things thereto belonging a common house, to bee called a Guest house for the lodging and entertaining of 50 persons in each upon their first arrivall'

Stores and supplies were frequently inadequate or of poor quality. The new settlers perished not in their hundreds or tens of hundreds but almost in their entirety. Based on the Virginia Company's own calculations during this period almost 4000 were transported of whom nearly 85% died. And these figures did not include the horrific final reckoning of the massacre of March 22nd 1621/22 when a further 350 were murdered. It was indeed a valley of death experience though not immediately for the Virginia Company but for most of the newly recruited settlers. Two of the necessary building blocks to create a tobacco industry were now in place. Innovation had opened up the possibilities and land grants had financed a workforce. What was now essential was entrepreneurial stimulus and this was provided by a wholesale reform of the pattern of ownership.

Ownership Reform

A major about turn in the Virginia Company's thinking was signalled by the announcement at the time of the departure of Sir Thomas Dale in April 1616, of an intention to grant significant tracts of land in lieu of cash as a dividend to the original investors and also to Ancient Planters who had been in Virginia during his time there. Though the grant to these early settlers would not be effective for three years the plan amounted to nothing short of a revolution. The most revolutionary aspect of the transformation in thinking was to grant the right for individuals to own plantations as private property and to grant property rights that would greatly extend the acreage under cultivation. In particular the cooperation and core key skills of these Ancient Planters, the 149 or so surviving Settlers who had arrived before April 1616, was secured by this grant of 100 acres for their own use. Richard and Isabel received 100 acres each. Isabel was one of only eight women to receive a grant in her own right. But the particular conditions attached to the grant of land to the Ancient Planters would not have been such good news. The Company clearly wanted to extract its last pound of flesh for there was to be a hiatus period with the workforce required to continue construction projects under Company authority for a further three years. So even now the Virginia Company had hesitated and further extended the brute force style rule. This has all the characteristics of a dead man's handle incentive, which constantly threatened to throttle back any immediate increment in vitality that the Virginia Company may have anticipated. But eventually the changes were promulgated in 1619:

'For the better establishing of a Commonwealth wherein order was taken for the removing of all those grievances which formerly were suffered ... and publishing a Proclamation that all those resident before the departure of Sir Thomas Dale should be freed and acquitted from such publique service and Cruell laws by which they had so longe been Governed'

'And that for all such planters as were brought thither at the Company's charge to Inhabit there before the coming away of the said Sir Thomas Dale after the time of their service to the Company on the Common land agreed shall be expired there shall be set out 100 acres of land for each of their Personal adventure to be held by their Heirs and assigns forever paying for every fifty acres thee yearly free Rent of one shilling ...'

When it was introduced in 1618/19 the most significant consequence of the new approach was to change in the relationship of power between the Virginia Company and the colonists to one that allowed entrepreneurial advance. The chance of gaining their freedom from the bonds of the Company had first been advanced by Sir Thomas Gates in 1614, to which was added a proposal for a dividend of land in April 1616. It is possible to imagine Richard and Isabel's mind set turmoil when they first heard of the proposal, for if it came to fruition it implied a most profound change in their lives. They must have felt that they were caught up in whirlwind situation, playing for the highest of stakes over which they had virtually no influence for they just had to wait and see how things turned out. Each day must have been a cliff hanging and nail biting experience caught between a breathless anticipation of a new life if things went right and a horrendous apprehension if the offer proved to be another false dawn. A saving grace was that from 1616/17 the dilemma was played out under the more benign colony leadership of George Yeardley.

It was the distance between the two possible outcomes that would have been so frightening. If the offer of a very significant dividend of land after another three years proved to be a broken promise Richard and Isabel would be plunged back into a waged servitude with little hope. But if the proposals became reality in 1619 then their manumission would not only mean that as Ancient Planters they would at last regain their personal freedom but also that they would each acquire a significant tract of privately owned land to farm on their own account with the prospect of exploiting the very profitable tobacco crop. When their wildest dreams were eventually realised the devout Richard and Isabel may well have reflected that it was no more than Jesus had promised in his Sermon on the Mount 'blessed are the meek for they shall inherit the earth'.

When the dramatic organisational change did come to pass in 1619 the next few years must have been very exciting times for they covered the crucial period when the fortunes of Richard and Isabel and his family, together with the surviving Ancient Planters were completely transformed from waged indentured servants to plantation owners with a prosperous future. The forced sea change in the thinking of the Virginia Company had led directly to the introduction of the private ownership of property. The Virginia Company, forced to abandon corporate responsibility for initiating progress, had unleashed the powerful force of individualism and self interest. But the introduction of excessive libertarianism was not accompanied by a coherent, intellectual road map. The new developments ensured that there would now be an atomistic fragmented approach to settlement with plantation owners focussing on their own patch with little or no coordination on the wider and deeper societal issues. For all its woeful inadequacies the Virginia Company had exhibited a limited broader perspective which embraced issues such as securing territory, establishing a working relationship with the local tribes, setting up a joint education facility, building infrastructure, regulating wages and keeping a weather eye out for the hated Spanish. But under the New Deal it remained the case that under reformed ownership there was still no plan of action to control ill considered emigration and so inappropriate workers continued to be shipped in at the wrong time of the year and left to fend for themselves in the woods until the growing and picking season started. There was still no attempt to isolate sickly new arrivals and a seeming acceptance of the consequent dreadful mortality.

The three year delay in implementing the land grant meant that Richard was not able to register and gain ownership until December 1620/21. Official records dating to May 1625 indicate that for a time Richard Pace had served as overseer of Captain William Powell's plantation on the south side of the James River He later left that post in order to develop his own land. Richard must have developed a close business relationship with William Powell for Virginia Land Office records show that he had a financial interest in a plantation that Captain William Powell intended to establish on the Chickahominy River. This would suggest that he was free from a commitment to the Virginia Company from 1619/20, after the three year additional service necessary to secure their 200 acres had been completed. During his time working as an overseer on the south bank, even though Richard did not have formal possession of his acreage, he would probably have had time to mark it out and even commence to build a dwelling and start to clear the land.

Individualism can breed greed and the new rules encouraged a land grab. The 1624 Colonial Map 'Curles in the James River' shows the extent of the take over with plantations strung out over many miles of river bank in a process that was bound to antagonise the indigenous Indians. There appears to have been no consideration of diplomacy but just a dangerous assumption that the local tribes would remain quiescent and resigned to losing their ancient lands. The decision to liberalize and decentralise enhanced the vacuum in oversight governance. Liberalisation without a strategic vision to create supporting structures, to allow some sense of fairness, justice or harmony with the potentially hostile resident tribes, or to attempt to ensure a strategy for security and defence would be the forces that would trigger a massacre.

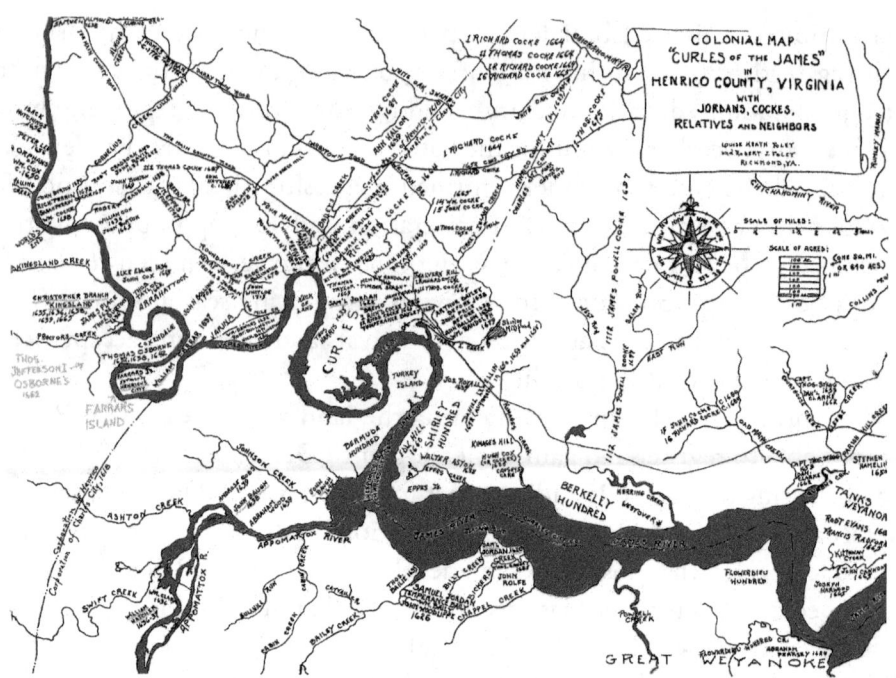

1624 map of the lower navigable portion of the James River

Private ownership, the availability of finance, and the innovative stimulus of the tobacco industry were mutually reinforcing. They were the key ingredients of the new stimulus. The Company allowed the dead hand of centralised control to be replaced by unleashing unrestrained individual enterprise and initiative. The ceding of power to the Ancient Planters was the action that shifted necessary conditions for commercial success to sufficient condition. Allowing this highly experienced and skilled body of men to take their own decisions and to work for themselves must have released an energy that amounted to a motivational explosion. This group was singular in that most were devout and not only had a self interest in success but also wished to do a good job to the highest standard in the sight of the Almighty. This new incentive impulse must have triggered a money-making fever flooding through the colony. The profits rolled in. The combined forces of individual initiative and radical organizational change which came about largely by chance, almost certainly saved the colony. By 1617 tobacco exports to England totaled 20,000 pounds. The next year shipments more than doubled. Tobacco was the frontier industry that allowed Virginia to take off into self sustained growth.

The history of the Virginia Company may be comprehended almost as a parody. It is as though having been granted certain formal monopoly privileges and status in the initial Patent, the colony seems to have become almost a private plaything of the leadership who initially focused on immediate self-enrichment. In 1609 there had been a last serious attempt to send significant resources but the expedition suffered appallingly bad luck from which it never really recovered. For five years under despotic rule the Company had tried to get an unruly and disaffected workforce to build a vainglorious infrastructure project and to expand the colony in a new location up-river. At first sight it may seem that now that the Virginia Company gave in and scattered its power to the wind like confetti. However this would have been far from the case, for the Company, who for years had jealously hoarded its power made a very reluctant retreat from authority. It had been able to use its only asset of land to buy itself out of its predicament but in so doing it released forces that inevitably eroded its power, undermined its position and forced it to adopt a more minimalist stance in relation to governance. The result of the vacuum and any societal perspective seems to have created almost a frontier town in a gold rush situation, although the gold was Virginia tobacco. As in England the development of a similar free enterprise environment would erode the power of the Guilds and allow rogue producers and traders to circumvent the statutory requirements of apprenticeships, wages, quality and price control, so in Virginia individual plantation owners disregarded the orders of the Virginia Company. The privatization of land, high consumer demand for the labour intensive tobacco crop and a shortage of manpower were the key driving forces. The new arrangements led to an enhanced tension between the colonists and Native Americans that would result in a massacre and also led to excessive mortality among the newly arrived workers who were often unable to cope with the harsh conditions, poor accommodation and a vulnerability to infection. Only after the catharsis of the massacre did the colony develop appropriate structures which introduced effective participative representation and justice. The criticisms in the Ancient Planters' Declaration are indicative of the shift of power. Though administrative decisions continued to be referred to London it may well be the case that this increasingly became more of a process of simply rubber stamping the de facto position on the ground in Virginia. For instance Richard and Isabel may have returned to Paces-Paines before formal permission to do so had actually been received. Democratic procedures were eventually introduced with local representation from each Plantation. Monthly Courts were set up in order to dispense justice, redress grievances and resolve petty disputes. These changes, coupled with an innovative and profitable tobacco industry, allowed the settlers to overcome the challenges to survival and develop a viable, prosperous colony.

In summary it is apparent that the Virginia Company displayed the arrogance of positional power. It is unlikely that even after the evidence of obvious failure of leadership became apparent they neither recognised, nor acknowledged, their shortcomings. They had failed to learn anything from the earlier experience. In 1616 roped together in the shipwreck of the Virginia Company, the Directors had taken a panicky decision to signal abandon ship. But the lifeboats were not serviceable and there were no life belts. In a shipwreck situation the order is eventually given 'every man for himself'. At that moment the ship's company ceases to exist as a community and becomes an assembly of individuals. Some will display altruism and heroism while others are prepared to put their foot in anyone's face to save their own lives. When the order to liberalize was given there was no mechanism to express the common interest and the settlement in Virginia disintegrated into a perspective of individualism. In retrospect it is possible to see a sharp contrast between this metaphorical shipwreck of the Virginia Company and the actual shipwreck of the *Sea Venture* on Bermuda in July 1609. In the latter case all survived and the reasons are not difficult to perceive. The leadership of an experienced seaman such as Sir George Somers was crucial. He had an immediate appreciation of the desperate situation and would have been aware that the odds were stacked against survival. He immediately formulated a strategy and a plan that gave a chance of survival. All on board worked in a coordinated way with skilled artisans such as carpenters being sent below to try to plug the leaks and buy time while all other able bodied men took their timed spell on the pumps. Aristocrats took their turn and pumped alongside men of lowly status. This was a concerted, directed effort with a common purpose understood by all. Contrast this with the Virginia Company leadership which in effect in 1616 gave the order to 'Abandon Ship'. In this situation there was no coordination or direction just chaos and a free for all. In the *Sea Venture* wreck, against the odds, all survived. In the case of the Virginia Company, arguably with much better odds, because of a lack of leadership, experience, strategy, plan and common sense, 85% of new emigrants perished. By 1616 the Virginia Company had required the kiss of life which was administered almost by default, but in so doing it also gave the kiss of death to many. The Company did eventually perish as well for it was wound up in 1624 when a royal inquiry led to the revocation of the Company's charter and its dissolution.

So the Virginia Company may be justifiably castigated for many of its actions. But there is one consequence of its legacy, even though it may have been instigated in desperation, which is so awesome and that would prove to be so transformational, that it dwarfs all these failings. The ideology or prevailing narrative under the regime of Sir Thomas Dale was that you toiled as you were ordered and conformed to the laid down code or you suffered severe punishment, even death, for disobedience. It was in effect a waged semi-feudal system. It was the introduction of the head right system and particularly the grant of land rights that re-purposed the colony of Virginia. The same process was to underpin the agricultural revolution in England with the enclosure movement and the industrial revolution that followed. The changes introduced a very different narrative. It demonstrated the value of private property and decentralized decision making as requirements for economic efficiency and political liberty. It unleashed the powerful forces of self interest and innovation. The two dominant societal objectives of the time were striving for military glory through conquest, and achieving eternal bliss through serving God. Now these were in competition with a new objective – the pursuit of personal wealth and to make society richer through economic growth. The shift in ideology signalled a new beginning which would have momentous consequences in the future. What came to be called Capitalism would prove to be more successful than any rival economic system. It presaged what America would become.

CHAPTER FIVE

Richard's Return and the Marmaduke Maydes

5

A KEY QUESTION to answer is why Richard and Isabel returned to England in the summer of 1621/22 and then chose to return to what until recently had been a hell hole. When they had sailed off as a newly married couple ten years earlier they were setting off, if not into the unknown, into the little known. They would have been optimistic and even in danger of wearing rose coloured spectacles. Immediately before they departed Richard was a jobbing journeyman carpenter eking out a living in Wapping, perhaps in the shipyard or may be in the building trade. Within the first year they underwent a baptism of fire which must have stripped away any illusions and exposed them to the reality of pioneer life. Assuming that they were on the *'Sea Venture',* then within a month they found themselves caught in an horrendous North Atlantic storm and after three days of trauma they had experienced a miraculous survival when they were wrecked on Bermuda. Richard helped to construct two barques on the island and they then journeyed the 800 miles on to Virginia. They arrived a year overdue to be met by only sixty emaciated, demented survivors from the happy band of their original fellow travellers. Perhaps Richard had chatted to young Jane on the quayside when the 3rd Supply fleet sailed. He would surely have learnt of the cannibalism that had taken place during the starving time. Her remains with clear evidence of this practice have recently been excavated at Jamestown.

For the next several years Richard and Isabel had slaved for their employers under a despotic regime as they hacked out a living for themselves and their son George from virgin territory. Now in the better times, as a condition of the transformative arrangements Richard and Isabel had each been granted 100 acres and had selected the land on the other side of the James River directly opposite Jamestown. They had now gained a land dividend and were free of obligation to the Virginia Company so why did they risk the return to England? They would have faced the daunting prospect of a double crossing of the Atlantic with trepidation. The very fact that they did sail is surely a reflection of their stoicism and a willingness to again submit themselves to the purposes and charge of the Almighty. However they did choose to leave their 12 year old son behind.

The couple would have had nostalgic reasons to visit friends and family for a last time. On arrival in England almost certainly in early summer 1621, Richard and Isabel may have stayed with his cousin Robert Clawson now aged about. 40 A few years later Robert's son Richard Clawson, a mariner, was living at Pare Tree Alley, Wapping, so perhaps this was their first stop. There may have been discussions about a possible future marriage between their son and Ursula Clawson who would probably have been of a similar age to George. This may well be the reason that Richard did not claim the 50 acres for Ursula as a new immigrant worker. Her return was a private family matter. The couple would surely have visited his brother Thomas and his expanding family at Kingston-on-Thames. Richard must have felt some pride that he could report that when he returned he would own 500 acres of land which would have seemed a huge tract of land to his brother. Richard and Isabel must also have seen young Thomas Rolfe now aged 6 so they could report back to his father on his progress.

The couple would also have visited their old friends and family in the vicinity of St James Clerkenwell and reflect with them on the manner in which their prayers had been answered. Richard and his wife would have been venerated and viewed as living proof of God's protection, for they had literally come back from the dead. In the autumn of 1610 when news eventually arrived of their survival they had been presumed dead for a year. It was shortly after these glad tidings that George Davison named his twins Richard and Isabel.

It is likely that gaining an additional 300 acres would have been a less pressing consideration underpinning their return, especially bearing in mind that George did not eventually register this in his name until 1628. The additional land was probably a longer term objective. A condition attached to the grant of a Dividend was that the land should be cultivated or 'planted' within two years. Richard would have had enough on his plate trying to get the original 200 acres cleared let alone an additional 300 acres. A deeper reason for wanting to secure this particular 300 acres may have been a collective decision taken with his immediate neighbours. If they were not able to secure this additional land on the fertile south bank of the James River, then the integrity of their contiguous holdings might be violated by a London capitalist who could finance some new workers and claim the land. These Ancient Planters had invested a good part of their lives in carving out a future and enduring much hardship and they could have resented the possibility of a 'cuckoo in the nest' in close proximity, who might not share their devout values. The prospect might have offended their sense of natural justice.

The most convincing primary reason for Richard to return would have been the urgent need to recruit sufficient workers to farm the land for the labour intensive, but lucrative, tobacco crop. The quality of any workforce available for hire, if indeed there was any surplus, was likely to be poor. By returning to England Richard would be able to select a reliable workforce from a known pool of friends and acquaintances or by personal recommendation. The cost of transport was possibly £6 per head but in addition there would be fitting out the recruit with suitable clothing and equipment as well as provisions for the voyage. Richard may also have offered a financial inducement as an incentive to sail for Virginia. The total cost may have possibly been in the region of £150 and this must have been a very considerable sum to find. Richard may have been able to have saved this over the years from his wages from the Virginia Company or possibly paid it by previous shipments of tobacco he had managed to make. However he certainly had managed to accumulate, or perhaps borrow sufficient funds to achieve his mission. The Ancient Planters at Paces-Paines may even have taken a cooperative decision to jointly finance the recruitment of the workers and Richard was the chosen agent to return and secure the land.

The evidence indicates that he returned to his old haunts around Clerkenwell and to his old friends and associates in order to select a quality band of reliable and trustworthy helpers. They would also have to conclude the administrative arrangements with the Virginia Company in London and to negotiate and secure passage for the entourage to Jamestown. So Richard must have spent a good deal of time in Clerkenwell in the vicinity of St James for it was from this area that he almost certainly gathered his recruits. Richard would have been unlikely to have accepted candidates without some form of prior knowledge or reliable character reference and a belief that they shared the religious commitment and attitudes embedded in the Virginia community.

Richard and Isabel selected the following six people to return with them:

1. Lewis Bayley
2. John Skinner
3. Bennet Bulle
4. Roger Macher
5. Richard Irnest
6. Ann Mason

Any information on the location of any of the six would prove useful in identifying their residence and family background. Initially indicative information on three of the six was found:

1. Lewis Bayley. A son Thomas Bayley, parents Lewis and Elizabeth Bayley, was christened in St Andrew, Holborn, London on 16th June 1620. It is therefore probable that Lewis Bayley left a wife and year old son when he set out for Virginia. It was not unusual for a wife to follow a husband to Virginia a year or more later. Lewis Bayley survived and appears in the later Muster.

2. Ann Mason, daughter of Richard Mason christened in St Andrew, Holborn, London on 2nd May 1610. If this is the same Ann Mason she would have been only 11 when she sailed for Jamestown and presumably recruited primarily as a house maid. Earlier genealogical research has shown that the Mason and Pace family had a long association. John Rolfe's mother's maiden name was Dorothea Mason and it is possible Richard Pace was his distant relative.

3. John Skinner. A John Skinner son of Thomas Skinner was christened at St Katherine Coleman church in the City of London 12th July 1601. The Muster at Neck of Land near James City, 1624/5 shows Skinner John, 1621 voyage aged 24, servant to Phetti Worthal. *(The Pace Society Data Base records Sara Pase marrying Robert Browne who was possibly born at Kingston-on-Thames at this church in 1598)*

No trace was found concerning the family origins of the other three new colonists Bennett Bulle, Roger Macher, and Richard Irnest, excepting one suitable profile, a daughter Margaret Bull born 8th December 1608, and a son Richard Bull born 7th October 1610 in London. Supporting evidence that Richard recruited from around Clerkenwell is that the names Macher, Bayley and Mason of some of the colonists who accompanied him to Jamestown were common in the congregation of St James Clerkenwell. The christening record 1551 – 1608, (no entries June 1552 – June 13 1560), include six entries namely, Mather, Mayor, Mawger, (Maugere), Mayer and Maior which might be alternatives for the name Macher; four entries for the name Mason who comprise the progeny of Hugh, James, John and Gowen Mason, and nine children of Robert, John, Richard, and William Bayley, (Bailey, Baylie).

So on their return in August 1621 it appears that Richard and Isabel recruited the new settlers from among their relatives, friends and the congregation of St James, or who were known to them through being resident in the immediate vicinity, such as the adjoining Parish of St Andrews Holborn, where as mentioned above, an Ann Mason and a child of Lewis Bayley were christened. All would almost certainly have shared a common religious conviction.

Also aboard the *'Marmaduke'* when Richard and Isabel sailed back were the first batch of young women recruited as potential colony brides. On 16th July 1621 Henry Wriothesley, 3rd Earl of Southampton, a member of the Board of the Virginia Company instigated a Subscription List to finance the sending of Maids to Virginia:

> *'Whereas by longe experience we have founde that the Mynds of our people in Virginia are much dejected and ther hartes enflamed with a desire to return to England only through the wants of the Comforts of Marriage without which God saw that Man could not live contentedlie no not in Paradize: ...*
>
> *We therefore judging itt a Christian charitie to relieve the disconsolate minds of our people ther and a speciall advancement to the Plantation to tye and roote the Planters myndes to Virginia by the bonds of wives and Children... in sending of young, handsome and honestlie educated Maides to Virginia: Ther to be disposed in Marriage to the most honest industrious ...*

Each share or 'adventure' was for £8 and 36 subscibers put up the money. Most purchased a single share but the Earl of Southampton put up £48 and Edwyn Sandys £40. The scribe Tristran Conyam had to laboriously write out 36 times:

> *'I am contented to adventure upon the aforesaid conditions ye some of –'.*
> *This was followed by the name of the signatory and the sum promised.*

The *'Marmaduke'* was a vessel of medium size of 100 tons that could carry 80 people which, after allowance for the crew, possibly included more than sixty settlers. From various sources it is possible to know the names of nearly half of them. Richard and Isabel and the six they had recruited, together with the thirteen maids listed below comprised 21 of those on board. In addition information from subsequent Musters in Virginia indicate the following also sailed on the voyage:

Ann Baly (wife of Nicholas); Fiona Chamberlain; Ann Doughtie (wife of Thomas); Katherine Fisher (wife of Robert); Andria Harries (wife of Thomas); Wm Querke aged 30 (servant to Francis Mason); Thomas Worthall (servant to Francis Mason); Ambrose – (servant to Henry Westwood).

The manifest of the Maids shipped out on the *'Marmaduke'* is shown below:

b) The names of the maydes sente in the Marmaduke bounde for Virginia An[n]o 1621. August

1 Lettice King — aged 23 yeares borne at Newberry in Barkeshire her father and mother are deade shee hath a brother that is an Atturny in the Law dwelling at Newbury [Here Conyam takes over from Ferrar.] Her father was an husbandman. S[i]r William Udall is her Cosen removed shee hath an Unckell dwelling in the Charter howse named Edward Colton.

2 Allice Burges — Aged 28 borne at Linton in Cambridgsheire her ffather and Mother are dead, hee was a husbandman She hath two Bretherne one a husbandman dwellinge at Linton the other a Souldier, Shee served about three years sithence one Mr Collins a silkeweaver right over against WhiteChapple Church after shee served Mr Demer a goldsmith in Trynity Lane, Shee is skillfull in anie Countrie worke, She can brue, bake, and make Malte &c.

3 Catherine Finche	Aged 23: borne at Mardens Parish in Heriforde sheire her ffather and Mother are dead, She was brought by her Brother Mr Erasmus Finch dwellinge in the Strand who is the Kings Crosbowe maker, with whom shee was and is in service, Shee hath likewise two other Brothers Edward Finch locksmith dwellinge in St Clements Parish without Temple barr and John Finch Crosbowe maaker dwellinge in St Martins Lane in
4 Margarett Bordman	Aged 20: years borne at Bilton in Yorkesheire her ffather and Mother dead, S[i]r John Gypson of Yorksheire is her Uncle by the Motherside, Shee hath bene in service w[i]th Captayne Wood who giveth a good testimony of her and so doth Mr Fynch havinge long knowne her, Her Mistris Mrs Kilbancke [give deleted] Mr Recorders Coachmans wife giveth a good testimony of her
5 Ann Tanner	Aged 27 borne att Chemsforde in Essex her ffather Clement Tanner dwellinge in Chemsforde by profession a husband man, Shee can Spinn and sowe in blackworke Shee can brue, and bake, make butter and Cheese, and doe huswifery, Shee hath a Cozen named Thomas Tanner sadler dwellinge within Algate
6 Mary Ghibbs	Aged 20: A mayde borne in Cambridge towne her ffather was a smyth, Her Mother is a live & dwelleth at Detforde Mr [Lott Peere added by NF] is her Uncle by her Mother side w[i]th whome shee dwelleth, shee can make bone lace, Mr Barbor likewise knowes her

7 Jane Dier	aged 15 born in St Cathems, her ffather was a waterman her Mother is a live, her name Ellen Dyer and dwells in St Catherines. Shee was com[m]ended by her Mother and goes with her Consent
8 Ann Harmer	Aged 21: borne att Baldock in Hartfordsheire her ffather is a Gentleman. Shee hath five Brothers and two Sisters Mr Underell ye grocer is her Cosen and Mr Fartlow a grocer both dwellinge in Bucklersbury, She hath an Uncle by her Mother syde named Mr George Kyngston [altered by NF to Kympton] now dwellinge att Weston, Shee hath bene brought up with Mr Morgan a Seampster, and can doe all manner of works gold and silke
9 Susan Binx	Mayd Aged 20: borne in St Sepulchers Parish in Seacoale Lane, her ffather and Mother is alive her ffather is a wyer drawer, Shee hath three Sisters, One Mrs Gardiner a gentlewoman and widdow in the Strand is he[r] Aunte by her Mothers syde, Shee was in service w[i]th one Mr Edward Batten a dum[m]er thatt dwells att the Lower end of Bartholemew Lane / and before in other good services, Shee can worke white and black work and knytt
10 Audry Hoare	Mayd aged 19: borne att Alesburie in Buckenham sheir, Her ffather and Mother are alive, Her ffather a shoemaker, She hath two Sisters one wherof brought her whose name is Joane Childe, dwelling in the Blackfryers downe in the Lane near the Catherne wheall, Shee had a Brother Called Richard. Apprentice to a fustian dresser, Shee can doe plaine worke and black works and can make all manner of buttons, One Mr Thomas Biling a marchant is her first Cosen and one Mr George Blunden an upholster in Cornwall

11 Ann Jackson	borne in Salisburie, Her ffathers name is William Jackson hee is a gardiner, and dwelleth in Tuttle side in Westmynster neer to the Redd Lyon Her ffather brought her, and her Brother John Jackson goeth for Martins hundred in Virginea

[Here NF takes over again from TC]

12 An Buergen	shee was shipped at the Isle of Wight by Mr Robert Newland in the roome of Mrs Joane Flechar whoe was turned back from thence.
13 Ursula CLawson	A mayde aged Kinswoman to Richard Pace an olde Planter in Virginia whoe hath given his bonde to pay for her passadge and other Chardges. shee wente in the compagny of the sayde Richard Pace and his wyfe.

Of the 29 persons, including the maydes, definitely identified, 19 were women and 8 were men. On the long voyage across the Atlantic Richard and Isabel would have got to know them all intimately.

The initiative to establish a syndicate to raise finance to transport young women to Virginia led by the 3rd Earl of Southampton had two main aims. First to encourage the Planters to concentrate on developing the long term prospects for the colony by setting down roots rather than constantly pining for the home comforts that it was suggested only a wife and family could provide and secondly to make some money. Their approach had all the hallmarks of previous projects associated with the Virginia Company. The initiative was ill thought through, hastily implemented, inadequately funded, insufficiently supplied and had little regard for the well being of the participants. Even in such a cruel age it was still a cruel action and Richard clearly wanted no truck with it as far his young kinswoman Ursula Clawson was concerned. It led to the inevitable early death of the great majority of young women who undertook to sail to a new beginning in Virginia. The subscription process would have had considerable kudos as it was led by such an illustrious personage supported by other eminent people. The girls were all provided with a complete outfit of clothes and accessories. The project was presented as a legitimate, authentic opportunity and some young women from various locations, supported by their relatives, were clearly sufficiently enticed by the attractive offer to apply.

An analysis of the first twelve maids to depart on the *'Marmaduke'*, listed is revealing. They are a surprisingly homogenous group. It was likely to have been a difficult position to be a female who had not married and who would therefore be a drain on a limited family income. Almost all are aged between 20 and 25 with the obvious exceptions of a 16 year old girl and a woman of 27. They all come from a similar respectable, though for some of the girls perhaps an increasingly insecure family background. Both parents of half of the young women were dead and a common thread is perhaps a hint that they were in danger of becoming rather inconvenient female relatives who were facing lengthening odds in the marriage stakes, heading for maiden aunt status and so becoming a financial burden as another mouth to feed. So there may have been subtle and even well intentioned pressure from relatives to promote the opportunity on offer as well as pressure on the girl to view the opportunity as a seemingly attractive deal. Throughout this project the propaganda glossed the opportunity on offer, in contrast to sounding any warning about the harsh reality of pioneer life. The maids may well have been seduced by this but the reality was to prove somewhat different.

It is possible to imagine these innocent manipulated young women, unwanted or surplus to requirements in their own land, excitedly chattering among themselves as they boarded the *'Marmaduke'*. The women folk on board would possibly have been quartered together aft under the poop deck and distant from the rough crew living in the forepeak. Knowing what would become of them it is a haunting image. Isabel would have been an enormously important person as, apart from her husband, she was probably the only person on board with extensive experience of the life style and the people, particularly eligible bachelors, the girls were about to join. Isabel would have been a mother hen figure and a fountain of information. Isabel probably pointed out the spire of St George's Church on the starboard side as they sailed down the Thames where Pocahontas had been buried five years earlier. As they entered the chop of the English Channel some of the maids would certainly have been seasick and perhaps already doubting the wisdom of their decision to emigrate. The stop-over at Cowes, Isle of Wight, may well have been scheduled in order to take on stores and additional cargo. The Earl of Southampton also held the title of Captaine of the Isle of Wight and probably had a base here for it was from here that Earl de la Warr had departed in 1610. At Cowes the 25 year old widow Mrs Joanne Fletcher suddenly left the vessel and was replaced by An Buergen. The reason is unclear though her details may give some indication:

> *Joanne Fletcher widdowe aged 25, daughter of John Egerton gentleman, brother to Sir Ralph Egerton, Knight home at Morley House near Bridge Stafford in Cheshire, this is testified by Mr Gibson dwelling neare to the Three Nunns, without-Aldgate*

If the details are accurate then Joanne Fletcher was better connected than the other women and also she was the only one who had been previously married. On the two day voyage round to Cowes no doubt she would have closely questioned Isabel as to the real situation, the quality of life and potential marriage suitors in Virginia. She was probably shocked at obtaining a reality check from Isabel and as a result promptly decided to abandon ship, which given her social status, was sanctioned. It is noteworthy that Richard paid for the transportation of his young relative Ursula Clawson. This would almost certainly have exempted her from any commitment to the marriage market in Virginia. It may have been the case that he had Ursula in mind as a potential wife for his son George who were only related through maternal sister grandmothers. It seems she was classed as a passenger and so he would have been unable to claim 50 acres for financing a new emigrant but her status and freedom of action in Virginia would have been protected by his action. Later documentation as to his land holding claim appears to confirm this.

After what was probably a 40 day or so voyage the maids arrived in Jamestown in October 1621. The Virginia Company said that given the short time period and rush before departure they did not have enough time to provide supplies to support the girls but they said that they would send some by the next supply ship. There was no guest house accommodation arranged for the girls at Jamestown but the Company said that they hoped that married Planters would put the girls up until they found a partner. It seems therefore that there were no welcoming arrangements or mentoring in operation and the girls were vulnerable for when they arrived they would be dependent for subsistence and accommodation on the goodwill of the settlers.

There must have been some sort of pairing off process. Gentlemen settlers would have been first choice husbands followed by Planters who held a substantial acreage. The maids who followed this first shipment would have probably missed out on the best prospects. The girls had little bargaining power apart from their gender and their attractiveness. Whatever the process it was swift for, before the '*Marmaduke*' turned round for the return voyage and so possibly within two or three weeks, all the girls had been snapped up and for better or for worse, found themselves in the marital bed. The Virginia Company did acknowledge the possibility that true love might intrude into the marriage compact and recognised that they must allow for this. Consequently they insisted that any unofficial husband who married for love must also pay the charge that was levied on other suitors and that the debt should have precedence over all others. This charge initially set at 120lbs of tobacco leaf per girl but because of a reported loss of condition of the tobacco leaf this was immediately raised to 150lbs for the next batch of maids. The charge no doubt enabled the subscribers to recoup, and make a significant profit from their investment.

The maids may well have not had very high expectations of life in Virginia but the shock of the life style change must have been very significant. Some would have adapted better than others. There was an obvious discrepancy between demand and supply which the Virginia Company sought to exploit commercially. The transportation of the maids to Virginia amounted to a legal abduction. If the real interest was to populate Virginia then the Company could have financed the project from their own capital. But they sought to make a profit from the opportunity and set out a cunning, but not quite false, prospectus to meet the wishful aspirations of their intended audience. Truly the action of a wolf in sheep's clothing. The girls might well have been angry and dismayed when they discovered the reality of their situation and it is unlikely that they were overjoyed. The truth was that these girls had been duped and sold into a marriage bed for a profit with little support or any concern for their suitability for life in pioneer territory. Within two years almost certainly only two or three would still have been alive.

CHAPTER SIX

Massacre: Trust is the Mother of Deceit

6

RICHARD PACE WAS very lucky to awake alive in the early hours of March 22nd 1621/22. Twenty four hours previously his servant Chanco had been instructed to assassinate him but Chanco must subsequently have wrestled with his conscience for he said nothing to anyone about his murderous assignment. When Richard retired to bed Chanco must have still been nursing his dilemma. Perhaps it was only when he crept into the bedchamber and as his eyes grew accustomed to the darkness and he saw the recumbent forms of his intended victims lying there before him that his feelings of affection overcame his sense of obedience to his King. So it is possible that he made his choice only immediately before he woke Richard and revealed all, but whether or not that was so it was certainly a late decision. Richard, half asleep must have struggled to fully grasp the enormity of what he was being told as Chanco blurted out his confession. Chanco, torn between conflicting loyalties, would have been anxious to exorcise his feelings of guilt for he had harboured his dastardly intentions for at least 24 hours and one might imagine that Richard might have looked at him a little askance when he realized how late in the day he had decided to confess. The servant then told Richard how much he respected him for Richard had always treated him as a son.

Richard's first reaction on being told the news must have been incredulity and shock at learning of the full extent of the planned attack. The colonists had been at peace with the Powhatan Confederacy for over eight years and there were seemingly no indications of overt aggressive intent and so this had probably allowed the settlers to be lulled into a false sense of security. Richard's second reaction must have been the realization of the desperate vulnerability of his local community and their unpreparedness to repel the imminent attack. Because they had been so complacent there was unlikely to have been a standing plan in place to match the scale of the challenge they were about to face. Richard would have realized that, in concert with his neighbors, it would be vital to first cobble together an immediate local action plan and then to consider the possibility of getting a timely warning across the river to Jamestown.

Though the following reconstruction of events is based on conjecture it is likely to be a broadly accurate picture of what happened. His two servants known to have survived, John Skinner and Lewis Bayley, would probably have been accommodated nearby, together with any others from the *'Marmaduke'* voyage still alive at this time. They would no doubt have been dispatched to the other residents in the vicinity with instructions for them to come in haste with all their people and armaments to an assigned assembly point. This would probably have been Richard's dwelling for recent excavations have suggest a possible location with a brick chimney but presumably earth fast posts in a field close to the river. It may have been fortified with a stout defensive palisade.

Excavation by Nicholas Lukketti, Prinicipal Archaeologist, James River Institute of Archaeology Foundation

It must have taken a considerable time to rouse and gather the other settlers from possibly four separate locations in the immediate area and would unlikely to have totalled more than twenty five persons. From the later muster it is probable that they included:

Richard and Isabel Pace and their son George, John Skinner and Lewis Bayley
William Perry
Thos Gates and his wife Elizabeth
William Bedford
Francis Chapman
Pettiplace Close, Daniel Watkins and servant Martin Demon
Edward Smith
There would have been other employees attached to each household

A typical settler's house with defensive palisade close to the riverbank

By this time it must have been at least 2 am and the defensive arrangements and disposition of armaments would then have to have been arranged. If the group possessed a cannonade this would have greatly increased their chances of survival. The discussion must then have centered on getting a warning to James Fort and a decision as to who should carry the fearful news had to be taken based on an assessment of the dire situation faced by the defenders at Pace's Paines. There must have been a consideration of whether self preservation should be prioritized over giving a warning to others. Sending someone across the river to warn James Fort would be a hazardous journey and meant depleting their own force. The emergency situation was telescoped into a very short period and Richard must have known that the future of the Virginia colony was likely to be decided in the next few hours.

The framework in which the action would be played out would be determined by both fixed and variable factors. Certain aspects can be taken to be fixed features of the situation on that fatal night. Cosmic certainties such as the orbit of the earth round the sun and the phases of the moon with its influence on the tides, remains virtually unchanged today. Physical features may be also thought of as constants such as the contours of the land, the course of the James River and the topography of the terrain. The geographical locations of the straggling line of dispersed plantations along both banks of the River James and particularly the three miles of broad river crossing between Pace's Paines and James Fort were established facts. Another permanent feature might include the psychology of danger. This would encompass aspects such as feelings of terror when faced with an imminent life threatening situation, a fear of the eerie silence of the dark hours or even concern about the ease with which noise travels over water at night which would be particularly relevant when engaged in a clandestine activity.

A variable factor that needed to be considered would have been the particular phase of the moon with its influence on visibility and tide. The tidal flow was probably the most significant factor affecting Richard's ability to get a warning across to James Fort. The state of the tide would have been of crucial importance in timing his departure for the crossing and it is almost certain that Richard must have set off just before slack water, possibly around 4 am when the tide might have been on the turn. Otherwise a tidal flow, which could reach a maximum of two knots would make the crossing more difficult or even have carried him past Jamestown.

This hypothesis was later confirmed by the tidal prediction obtained from

Massacre: Trust is the Mother of Deceit

the National Ocean Service. It is dated in the modern Gregorian calendar which was adopted by Britain and the USA in 1752 and which 'lost' 10 days from the Julian calendar. To compensate for this Friday March 22nd 1621/22 becomes Friday April 1st 1622 in the modern format. It shows low water was at 9.48pm on the evening before the massacre around the time Richard probably retired to bed. High Water was fortuitously at 0356 am in the early morning. So almost certainly Richard rowed across the James River between 3.30 and 4.30 at the top of the tide, during the period of slack water.

Jamestown Wharf, James River, Virginia
Tide Predictions (High and Low Waters)   March, 1622
0.0000000
NOAA, National Ocean Service

Standard Time

Day	Time	Ht.	Time	Ht.	Time	Ht.	Time	Ht.
21 M	12.07am	L0.4	6.14am	H1.9	1.04pm	L0.6	6.44pm	H1.7
22 Tu	1.13am	L0.3	7.20am	H1.9	2.05pm	L0.4	7.50pm	H1.8
23 W	2.19am	L0.2	8.24am	H2.0	3.02pm	L0.2	8.51pm	H2.0
24 Th	3.22am	L-0.1	9.22am	H2.1	3.55pm	L-0.1	9.48pm	H2.2
25 F	4.20am	L-0.4	10.17am	H2.2	4.46pm	L-0.4	10.41pm	H2.4
26 Sa	5.15am	L-0.6	11.09am	H2.3	5.35pm	L-0.6	11.33pm	H2.5
27 Su	6.08am	L-0.7	11.59am	H2.3	6.23pm	L-0.7		
28 M	12.24am	H2.6	7.00am	L-0.7	12.49pm	H2.3	7.12pm	L-0.7
29 Tu	1.15am	H2.6	7.52am	L-0.6	1.39pm	H2.2	8.02pm	L-0.6
30 W	2.07am	H2.5	8.45am	L-0.4	2.30pm	H2.1	8.53pm	L-0.4
31 Th	3.01am	H2.4	9.39am	L-0.2	3.24pm	H2.0	9.48pm	L-0.2

April 1622

01 Fri 03.56am H

The second key factor beyond Richard's control during his crossing of the James River was the wind force and cloud cover. Although the prevailing weather conditions are not known it is possible to surmise what the situation may have been based on current weather reports for the area. The weather report for the same date nearly four centuries later gives an indication of what the conditions might have been on that early spring morning 1621/22. Over a six day period there is no cloud cover on five of the six days. The lowest early morning temperature is just above freezing. Another key factor would have been the direction and force of the wind especially if it was against the current which would have affected the surface condition of the river and possibly rendered the crossing hazardous. The wind speed is surprisingly volatile in this sample and varies from relatively light winds from the north to near gales from the SSW on two occasions on the Tuesday and Thursday nights. Gales of this magnitude might well have meant that a crossing impossible, or at best would have resulted in a very rough passage in choppy waters. At least if the wind had been from the south Richard would have been blown across to James Fort. On the early morning of March 22nd the moon rose in the SSE in the constellation of Sagittarius at an hour past midnight so as Richard rowed across, the moon would have been high in the sky shining in his face from the left. Interpolating from the 2016 weather reports for the same period in March it is possible to speculate that on the night Richard crossed, the morning would have had a chill with a clear sky, there might have been a light wind from the north and a three quarter waning moon positioned not far off the meridian which would have provided considerable ground visibility.

The James River Crossing by moonlight from the north bank.

If it was a clear night it would have been unfortunate for it would have bathed Richard in moonlight increasing his vulnerability as he rowed across. Given this possibility, the amount of cloud cover may have been an important factor in determining whether or not Richard was discovered.

Weather Prediction for Friday 25th March 2016

Mulberry Point, James River tide table for the next 7 days
Issued (local time): 7 am Monday 21 Mar 2016

Days 0-3 Weather Summary: Mostly dry. Cool air temperatures (max 19°C on Wed afternoon, min 2°C on Mon morning). Winds increasing (light winds from the NW on Mon night, near gales from the SSW by Tue night).

Days 4-6 Weather Summary: Moderate rain (total 11mm), heaviest on Thu night. Warm air temperatures (max 21°C on Thu afternoon, min 7°C on Fri morning). Winds decreasing (near gales from the SSW on Thu night, light winds from the N by Fri afternoon).

name	Monday 21			Tuesday 22			Wednesday 23			Thursday 24			Friday 25			Saturday 26		
	morning	afternoon	night	morning	afternoon	night	morning	afternoon	night	morning	afternoon	night	morning	afternoon	night	morning	afternoon	night
High Tide	10:52AM		11:08PM	11:31AM	11:47PM		12:07PM	12:23AM		12:42PM	12:57AM		1:16PM	1:31AM		1:50PM	2:05AM	
height (m)	0.76		0.75	0.77	0.77		0.77	0.79		0.76	0.79		0.75	0.75		0.75	0.78	
Low Tide		5:24PM	5:40AM		6:01PM	6:21AM		6:36PM	6:59AM		7:09PM	7:35AM		7:42PM	8:12AM		8:15PM	
height (m)		0.00	-0.02		-0.00	-0.02		0.00	-0.01		0.01	0.01		0.03	0.04		0.05	
Wind (km/h)	30	30	20	20	45	55	45	45	45	40	35	45	20	15	15	15	15	15

Today's sea temperature in Mulberry Point, James River is 4.8 °C (Statistics for 21 Mar 1981-2005 – mean: 10.4 max: 14.5 min: 7.0 °C)

Summary	some clouds	clear	clear	clear	clear	clear	clear	clear	clear	some clouds	rain shwrs	rain shwrs	clear	clear	clear	clear	clear	some clouds
Rain mm	-	-	-	-	-	-	-	-	-	-	8	3	-	-	-	-	-	-
High °C	5	10	9	8	14	12	12	17	17	18	21	19	9	13	11	10	12	14
Low °C	2	8	4	4	11	9	8	17	13	13	21	11	7	12	8	7	12	11
Chill °C	0	7	5	5	11	8	9	18	15	17	21	18	6	12	8	8	10	12
Sunrise	7:07	-	-	7:05	-	-	7:03	-	-	7:01	-	-	7:01	-	-	7:00	-	-
Sunset	-	7:19	-	-	7:20	-	-	7:21	-	-	7:22	-	-	7:22	-	-	7:23	-

Monday 21st – Wed 23rd:
Mostly dry. Cool air temperature, (max 19C on Wed afternoon, min 2C on Monday morning. Winds increasing, (light winds from the NW on Mon night, near gales from SSW by Tues night).

Thursday 24th – Sat 26th:
Moderate rain, (total 11mm), heaviest on Thursday night. Warm air temperatures, (max 21C on Thurs afternoon, min 7C on Fri morning). Winds decreasing, (near gales from the SSW on Thurs night, light winds from the N by Fri afternoon).

Phase of the Moon March 1621/22 *(Julian Calendar)*

New Moon

Full Moon

March 1622 (United States)

Sun	Mon	Tue	Wed	Thu	Fri	Sat
					1	2
3	4	5	6	7	8	9
10	11	12	13	14	15	16
17	18	19	20	21	22	23
24	25	26	27	28	29	30
31						

New Moon

March 22 1621/22
Waning quarter phase moonlight

Moon Phase Calendar

In addition to these fixed and variable factors, the dynamic of the interaction between the opposing forces would be conditioned by the number, deployment, and motivation of the attackers and the preparedness and armaments of the defending settlers. The evidence suggests that the Powhatan attackers were not vigilant during the night hours immediately before the attack. They apparently had no clue that one of their own might betray their plan and so possibly believed that there was no perceived need to maintain a watch or river patrol which might well have detected any preparations to carry a warning to others. Richard was selected as the person who should take the news to James Fort probably because he was the one to whom Chacun had passed the information. It may also have been the case that Chacun actually rowed across the river with Richard for he would then have been able to deliver a verbatim account of the plan and be further questioned by Governor Sir Francis Wyatt about the full extent of the proposed attack. He would also be an additional oarsman to help in the long row across and as the great betrayer he may not have been anxious to be around when his tribal brothers came calling in the morning. Richard's work boat may well have been designed to be used for shipping his tobacco crop out to trading ships lying in the river. If so it would have been strongly constructed no doubt by Richard himself, probably at least twenty feet long and able to carry a couple of hundredweight. It was unlikely to have been in daily use at this time of year and may have been laid up for the winter and hauled out adjacent to the slipway, lying upside down to prevent it being filled up with water during rain storms. The oars were probably stowed in the house over the winter months. The craft was therefore likely to be fairly heavily built rather than a lighter skiff and probably had a fitment to ship a mast designed to carry a lug sail when required.

As the settlers gathered in Richard's house close to the river there must have been a desperate concern that the attackers might be in the offing. Perhaps a small armed group of perhaps six crept down from the house under the moonlight carrying the oars and launched the boat to the bank. As they did they must have been subject to feelings of fear, deeply stamped on the ancient soul of humankind. If they were intercepted by a war party in the open then they were dead. It must have been an electrifying experience with nerves and senses tightened, stretched, and tuned as on few other occasions. Perhaps they wrapped the oars with canvas or wool where they rested in the crutch in order to provide a pad for silence. In the event they were undisturbed and they would have watched with relief as Richard splish-splashed away on his errand of mercy. It must have taken the best part of an hour for him to traverse the river. Whatever the details of his departure may have been, Richard eventually made it across, roused the guard and delivered the warning to the Governor. Jamestown Island had been chosen originally as the site of the first colony base for its defensive characteristics. A triangular palisade fort had been constructed with one side fronting the river to guard against the possibility of a water-borne assault, which also had a gate. Defensive towers mounted with ordnance would make a direct siege attack very dangerous and costly to life. The town had subsequently been extended and dwellings spread outside the fort walls and beyond the island itself onto the neck of land and spreading out on the mainland. There must have been sufficient time to bring everyone in the immediate area within the protective bastion. Once the warning had been received the attack on Jamestown was doomed to failure In the face of canon mounted on the guard towers, any assault, even a multi-pronged attack, would have been decimated by cube shot. The most authentic account of the events of the early morning of Friday 22nd March is likely to be derived from a contemporary record written by someone who probably talked directly from Richard soon after the events:

That the slaughter had been universall if God had not put it into the heart on an Indian belonging to one Perry, to disclose it, who living in the house of one Pace was urged by another Indian, his Brother, (who came the night before and lay with him), to kill Pace, (so commanded by their King as he declared), as he would kill Perry: telling further that by such an houre in the morning a number would come from divers places to finish the Execution who failed not at the time: Perries Indian rose out of his bed and reveales it to Pace that used him as a sonne: And thus the rest of the Colony that a warning given them by this meanes was saved:

Thousands of ours were saved by the means of one of them alone which was made a Christian: Blessed be to God forever, whose mercy endureth forever: who wrought this deliverance whereby their soules escaped even as a Bird out of the snare of the Fowler:

Pace, upon this discovery securing his house before day rowed over the river to James City, (in that place neare three miles in bredth), and gave notice thereof to the Governor by which meanes they were prevented there and at such other Plantations as was possible for timely intelligence to be given: for where they saw us standing upon our Guard at the sight of a Peece they all ranne away.

From the reports of what transpired and an assessment of what happened it is possible to discern something of Richard's character. He was certainly devout and a kind and considerate employer. He was altruistic and concerned for the fate of others in the community and put that ahead of the perilous situation of his own wife and son. But he must have been decisive and able to see beyond the immediate threat for he must have calculated that if Jamestown fell then it was probably inevitable that they would all perish as the attackers picked off remaining settlements at their leisure.

Other settlements were not as fortunate as James City. On the morning of the 22nd of March 1621/22, three days before New Year's Eve and exactly four weeks before Good Friday, the Powhatan Indian Confederacy launched a concerted and coordinated attack on thirty one of the English Settlements strung out either side of the James River in Virginia. The date chosen for the massacre attack may be have symbolic importance for March was unusual in that there was a new moon on the first and last day of the month. It may be significant that the attack took place close to the Spring Vernal Equinox when the sun crosses the celestial equator from south to north when night and day are of equal length, an event regarded in some cultures, as a time of rebirth. The Powhatans were skilled lunar timekeepers and tallied the moon on strings and notched sticks used to reckon such times as the moon of stags, the cow moon and the first and second moon of cohawks.

The Powhatan leadership must always have been concerned that the scale of colonization could be contained and that any increase should be on their terms. The first significant influx had come with the 250/300 arrivals on the 3rd Supply in the summer of 1609, but the subsequent reduction to 60 survivors following the starving time after being besieged, may have allayed Powhatan's fears. It would also have given some re-assurance to the tribes for they now had an historical precedent that seemingly proved they had a control mechanism to hand. The setbacks and lack of commercial success experienced by the Virginia Company had resulted in a fairly static population during the period 1607-1616, and this possibly acted as a veneer which concealed the potential for a significantly increased scale of future colonization. By 1618 this potential was realized as over the next three years, four thousand or so new settlers poured in. Metaphorically the Powhatan Confederacy glimpsed the face of the colonial tiger – and it was not smiling. The tactic of control by besiegement was replaced by one of annihilation.

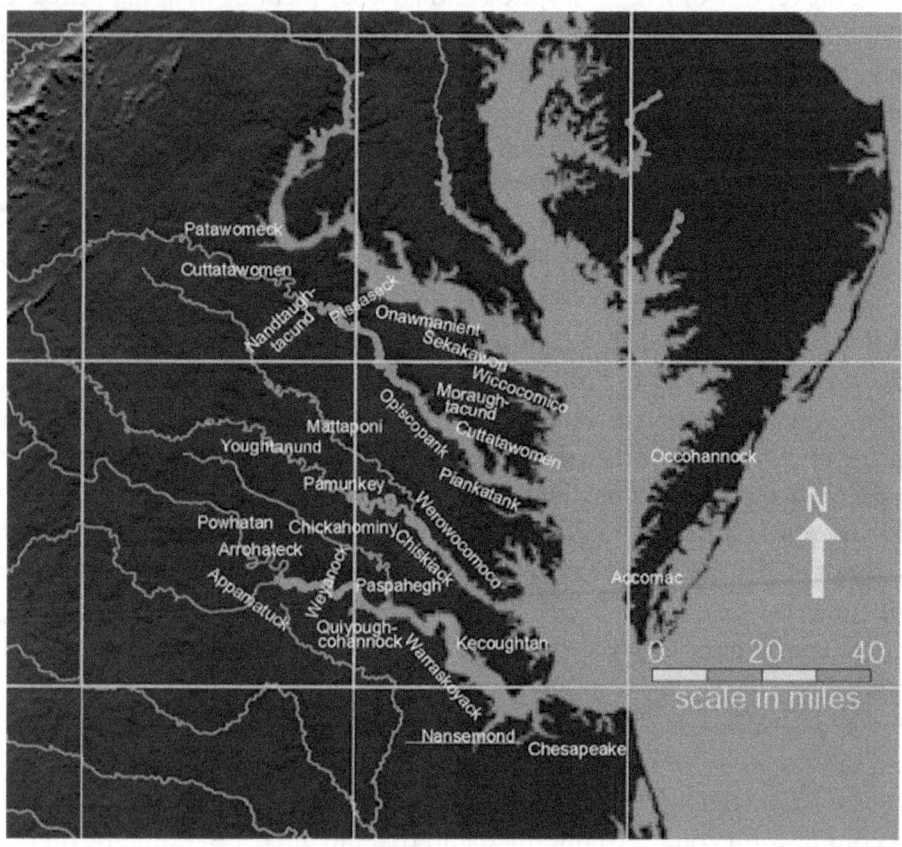

Jamestown Tribal Territories
(Reproduced by kind permission of Virtual Jamestown, VCDH. Univ. of Virginia)

The map shows the location of the various elements of the Confederacy which comprised over 30 Algonquian speaking Native American tribes spreading out from Chesapeake Bay Virginia. The plan almost certainly must have involved delegated responsibility to each local tribal chief or 'weroance' who paid homage to the Powhatan ruler of the Confederacy.

Though the trigger for the massacre has been reported to have been the murder of a close adviser to the Powhatan leader, the underlying reason must surely have been alarm at this huge increase in the number of new arrivals and the seizure of territory as the settlers acquired their Dividend of land along the James River. The term Jamestown Massacre conjures up an image of indiscriminate random slaughter by disorganized, rampaging savages as they roamed the countryside intent on killing any victim settler they happened to encounter. This must have been far from the case. This was not a whim attack, or a casual killing spree by a handful of marauding Native American tribes, but a considered, concerted and well coordinated plan to annihilate the colonial settlement in Virginia. The massacre could only have been based on long term strategic thinking that allowed the implementation of a sophisticated battle plan with clear objectives, pre-identified targets backed by established, well understood, clever tactics. The attackers would have calculated that the opposition would be ill prepared, poorly armed, complacent, off their guard and therefore very vulnerable. Opechancanough had become the paramount Chief of the Powhatan Confederacy in 1618 after the death of his brother. He was an experienced and wily leader and the detailed organization of the attack clearly reflected this.

Though the strategic battlefield area which encompassed a considerable distance along both banks of the navigable James River and which penetrated into the hinterland, comprised an area of more than a hundred square miles, the actual operational area was much smaller. It included just over thirty identified specific targets of which four were significant townships. In reality the eighty mile stretch of river bank with about eighty discrete settlement sites was then reduced to no more than a handful of targeted operational locations. It is apparent from the detailed treacherous role that Chacun had been instructed to carry out at Pace's Paines, that the planning must have been very precise. Though the assault on Jamestown failed in general the tactics used by Openchancanough the Powhatan tribal leader were brilliantly conceived and meticulously planned and based on a ploy designed to deceive. Openchancanough must have discussed each of the intended targets in the light of the detailed knowledge provided by the local leader, including the size of the raiding party, the sequence of the attacks and any possible obstacles to success. The delegation process would have had the advantage not only that the local tribe attackers possessed an intimate knowledge of the terrain and patterns of behaviour of the settlers but that particular tribes were killing in order to regain their dispossessed territory. Humankind seems to have an inbuilt emotional attachment to land and this may have heightened motivation and driven a greater frenzy. Each raiding party would have had a laid down itinerary for the particular attacks and been well aware of the best ambush points and the most appropriate landing beach. The leader of the Powhatan Confederacy would have laid down the precise timing of each attack. He would have established what had to be achieved at each location and crucially the tactics of subterfuge and surprise to be employed and maintained for as long as possible. He seems to have refrained from an attack on Elizabeth City, perhaps because of its more remote location or possibly in order that it could be retained as a trading outpost by a small number of surviving settlers under the tight control of the Powhatan Confederacy.

Opechancanough was playing for very high stakes in seeking to encourage the Powhatan tribes to regain the land of their fathers from the interlopers but he must have reckoned that he had the balance of advantage. The settlers had very little going for them given their situation. Apart from some carefully chosen defendable locations such as Jamestown and Henrico Island, many settlements were scattered along the river probably possessing some inadequate, ancient ordnance with perhaps a homestead with a palisade to act as a strong point. In the event these inadequate defenses proved virtually useless in the face of the fiendishly clever tactics of subterfuge and surprise. In contrast, when given a few hours prior knowledge, Richard and his near neighbours were obviously able to swiftly prepare a defensive plan at Pace's Paines that saved their lives. The same applies to Jamestown where early information about the intended assault saved all there. Where there was no warning there was carnage. Had Jamestown fallen then possibly in total two thirds of the population may have died. This would probably have amounted to a critical mass which would almost certainly have brought about the end of the Virginia Settlement in any meaningful form.

From the mouth of the estuary in Chesapeake Bay the James River runs for 100 miles, generally in an east-west direction before it turns due south for 8 miles flowing parallel to, and just east of the head of navigation known as the Fall Line, before resuming its previous direction. At Jamestown the river is still deep and three miles wide but this is the end of what may be thought of as estuarial waters. The expanse of water rather resembles a lozenge shaped lake, which would be liable to significant choppy conditions in high winds. It then becomes much narrower and tortuous as it flows on. It was the dispersion and isolation that made the individual plantations so vulnerable to attack from either land or from the river. This 100 mile stretch of river and the plantations strung out along, and penetrating the near interior of the river, comprised the extremities of the chosen battlefield for the guerilla and subterfuge tactics that were used by the Powhatan Confederacy. Four main towns or incorporations lay along the river. Kecoughton, later renamed Elizabeth City, which lay nearest to the river mouth, was not attacked. Jamestown was the second incorporation and the main centre of population lying on the north bank of the river and some 30 miles upstream from the estuary. The assault on Jamestown, the main centre of population, would have been the central plank of the attack and would have presented much more of a success than wiping out smaller, more isolated groups.

There seems to have been two main thrusts for the attack at the lower and upper reaches of the navigable James River. At the lower part the greatest loss of life in a single location took place at Martins Hundred which comprised 80,000 acres on the north shore, just 7 miles downstream from James City which had been settled by 250 arrivals in 1619. The administrative centre was Wolstenholme Town, a fortified settlement of rough cabins. Nearly 80 settlers perished at this location including a number of women and children although in the records the Wolstenholme dead are not enumerated separately. Over fifteen women were taken into captivity by the Indians. The Plantation had a extensive river frontage so it would have been particularly vulnerable to an assault by a fleet of canoes. There was clearly not enough time for Jamestown to send a warning by boat, even though the tide would now be ebbing. Perhaps having been frustrated in their intended assault on Jamestown there were more attackers in the vicinity to hit Martins Hundred. After the massacre Martins Hundred was abandoned and only resettled a year or two later. Across the river and downstream of Hog Island, Bennetts Plantation, on what became called the Isle of Wight, was decimated where 54 were slain. The new settlers had only arrived two weeks before and would have been disorganized and have had little time to establish any defenses. Sixty miles up-river Henricus Town, on Farrar Island, with three main streets and a church, which had been constructed from 1610 under the direction of Sir Thomas Dale, was attacked. The site location had been chosen for similar defensive characteristics to Jamestown, being a virtual island with a 7 mile river frontage. The seeming security offered by this narrow neck with the river acting as a defensive moat proved only too vulnerable to the Powhatan tactics. It is probable that the woodcut by Matthaeus Merian 1628 comes close to depicting what actually happened here.

It has been reported that, certainly in some cases, the Indians came into a settlement unarmed, on the pretence of a usual trading practice seeking to exchange food such as turkeys and pigeons for trinkets copper and beads and other paraphernalia. This was very possibly the tactic used at Henrico Island. The expected trading session on that Friday morning would have allowed a substantial number of Native Americans to infiltrate Henrico without arousing suspicion. They simply walked in through the narrow defensible neck of land entrance and were no doubt welcomed in friendship.

At a signal the slaughter began as the visitors suddenly turned on their hosts, seizing whatever artifact or proxy weapon was to hand. Synchronised support would then have arrived from a flotilla of canoes lying just offshore and bringing significant reinforcements. As a surprise strategy it was brilliantly successful, a blitzkrieg onslaught against which there could be no immediate defense. The killing was indiscriminate Not only were the adults slaughtered but women, some of whom were with child, children and babes. The terror must have been unimaginable. The thud of clubbing and stabbing, the screams and the war hoops fading and then just a pathetic whimpering and eventually a dreadful silence as the ritualistic dismemberment of the dead and the dying was carried out. The death toll of around thirty at Henricus Island seems surprisingly small but it would have taken time to stab and bludgeon people to death with inadequate weapons. Some may have had time to run for their lives and hide up. Once all in the immediate vicinity had been dispatched and the gruesome rituals performed the attackers left, perhaps anxious to move on to other locations before they were alerted which would still allow them to use a similar subterfuge ploy. In other settlements particularly those without river frontage, a different strategy was probably used. The attackers would have been familiar with the layout, the daily routine and the pattern of work in a particular plantation. They would have known where the men folk were located at a particular time and could have employed this local knowledge to their advantage in timing an attack.

The map below taken from the Gutenberg Project shows the navigable stretch of the James River to the Fall line and gives an excellent perspective of the complexity of the area of settlement. It indicates the location of some of the main attacks such as at Falling Creek which empties into the James River. It was this tributary that provided the water power for the iron works. Here Captaine Berkeley's Plantation had 27 slain and to the south 3 miles away Thomas Sheffield's Plantation reported 13 dead. Just two miles from Henrico City, 17 of the staff at the newly established College were slain.

Massacre: Trust is the Mother of Deceit

Plantation deaths along the James River 1624 (Courtesy Gutenberg Project)

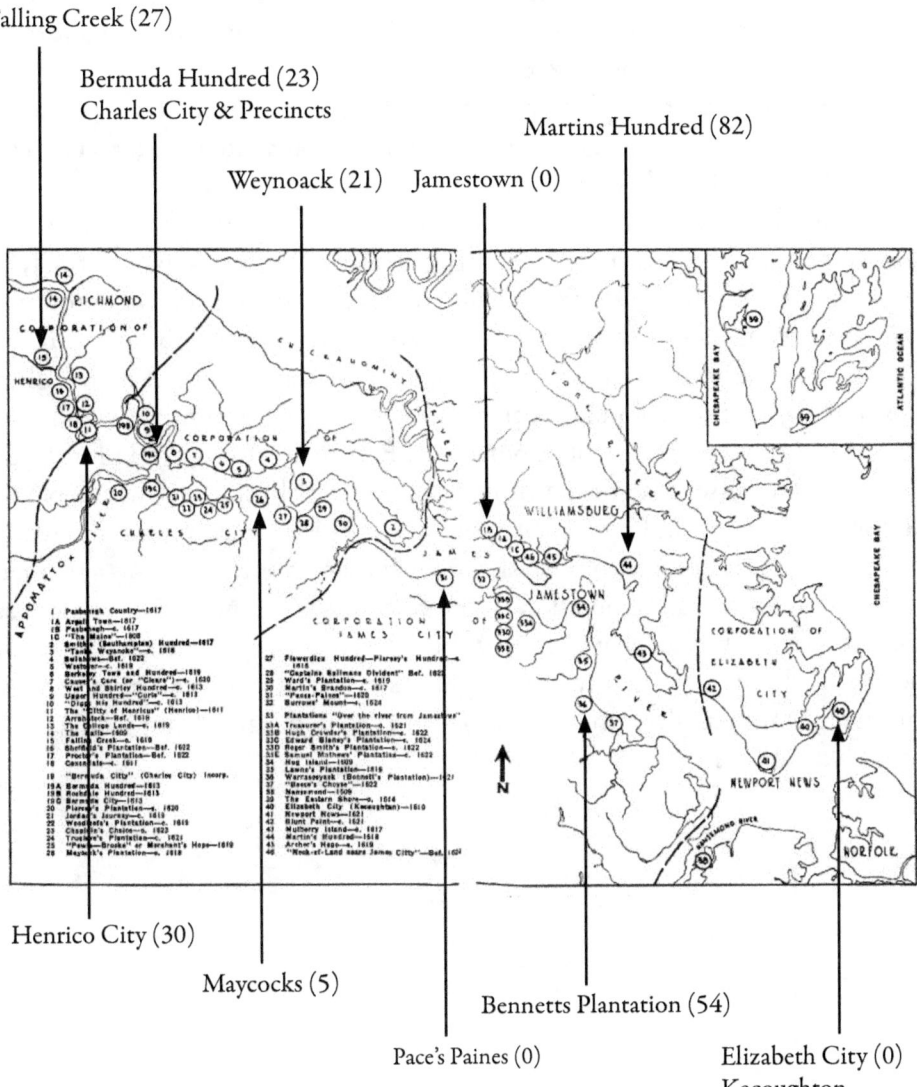

Falling Creek (27)

Bermuda Hundred (23)
Charles City & Precincts

Weynoack (21)   Jamestown (0)

Martins Hundred (82)

Henrico City (30)

Maycocks (5)

Bennetts Plantation (54)

Pace's Paines (0)

Elizabeth City (0)
Kecoughton

In the vicinity of Charles City, previously called Bermuda Hundred, there were a number of separate attacks. These included Berkeley Hundred which had been settled from Bristol in 1619 where 11 died, only two of whom had arrived on the original voyage. A few more died at Westover about a mile from Berkeley-Hundred. At Flowerdieu Hundred at Sir George Yeardleys Plantation 6 died, and on the other side of the river opposite Flowerdieu Hundred 7 people, including 2 women, were killed.

When Opechancanough reckoned up the outcome of the Massacre plan he must have reflected that he had decimated most of his intended targets except Pace's Paines and the key stronghold of Jamestown. These were the only two locations where there had been a prior warning His failure to take Jamestown meant that he had led a failed enterprise. Partial failure was eventually to prove to be total failure. But given his mind-set and cultural background it is unlikely that he grasped that this was an all or nothing situation and that as a result he and his people would come to lose their heritage. Although he continued to oppose colonial expansion over the next years until 1644, this was the crucial missed window of opportunity. Eventually it was this failure, more than anything else, that would bring a terrible retribution on his people for the sneak attack would be used as moral justification for a process that would ensure that in the long term the indigenous Indians would end up as a remnant population. The total population of the colony was around 1275 and at least 345 perished. The fact that over 800 survived was solely due Richard's heroic action. Richard's own family and neighbors survived at Pace's-Paine and it is probable that here given the absence of an expected signal from Chanco, the attack was aborted.

CHAPTER SEVEN

Aftermath and Rebirth

7

THE CONTEXT OF the massacre and the effect on the settlers can perhaps best be understood in terms of the sharp contrast between the good times at last being enjoyed by the colonists immediately before the attack, and the situation in the aftermath. In the years before the massacre there had clearly been a huge improvement in the morale of the settlers compared with the bad old days under the regime of Sir Thomas Dale. The Ancient Planters were in good heart for they described these times in glowing terms:

> *The plenty of these times likewise was such that all men generally were sufficiently furnished with Corn and many also had plenty Cattle, Swine and poultrie and other good provisions to nourish them*'

The feeling of contentment was further increased with the arrival of the new Governor confirming the grant of new freedoms that had been promised earlier:

> *'In October 1621 Arived Sir Francis Wyatt Knight with Commission to be Governor and Captain General of Virginia: he ratified and confirmed all the aforementioned liberties, freedoms and priveledges to our great happiness and content: the Countrie allso flourished and increased in her former proceedings as iron works, plantinge of Vines and Mulberye for silke'*

Life can be uncertain but to go from this euphoria to the depth of despair after the attack just five months later must have been a terrible shock. It is worth trying to imagine the mind-set of the settlers on the morning after the massacre and over the next few days as the full horror and extent of the attack filtered back as the settlers abandoned their plantations and crowded into the safety of Jamestown. In the end only eight of the original eighty settlements remained occupied. Jamestown must have been like a refugee camp with the settlers arriving with the bare essentials of all they could carry.

The sheer scale of the deaths must have been a terrible emotional shock and difficult to comprehend. Descriptions of the actual events at particular locations would have been recounted and listened to with disbelief. There would have been frantic efforts to find out who had died and who had escaped, with tales told of miraculous survival. For instance in one situation a settler seeing a murder taking place, had seized a shovel and beaten off the attacker and then, having alerted others, together they had managed to drive off the assault, which the Native Americans had not pressed home. The settlement was still a small community and although spread in numerous locations over fifty miles up and down the James River, people would have either known those killed quite well or at least known of, all of those slaughtered. Even so, hearing of a death of an acquaintance would have provoked a very different emotional response than hearing of the death of a member of the family or an intimate friend.

For the survivors of Martins Hundred, just seven miles downstream of Jamestown where the majority of the inhabitants had been killed and some of the deaths witnessed, the trauma and sense of guilt that they had somehow survived, would have been even sharper for had a warning got through, the township might have been saved. There would have been horror expressed about the recent shipload of Puritan arrivals at Bennetts Warrosquoake Plantation who had lost nearly half of of their fellow colonists. The *'Sea Flower'* had brought 120 new settlers in February 1621/2 of whom 53 were killed. They had been there less than a month and so would have had little time to have established even rudimentary defences. The people at Reverend Maycock's isolated plantation several miles upriver from Pace's Paines would have had little chance of survival. The clergyman had come out to Virginia with his wife in 1618 and had chosen to develop a plantation. The family holding was wiped out but afterwards their infant daughter Sara was discovered apparently deliberately hidden by the parents in the face of the imminent attack in which they both were murdered. Elsewhere there were tales of heroic rearguard actions that repelled surprise attacks.

But for a society on the edge of survival there would have been little time or even an appetite for protracted mourning. It is probable that after the initial catharsis people would not have wanted to revisit memories that still haunted them. The settlers would have had had little choice but to discard the past and get on with life as best they could. And times were hard, for supplies were short and the next harvest poor. Living from hand to mouth again inevitably focused their minds on the present. No doubt the colonists were initially fearful of a renewed attack and cowed by the experience they had endured. Over time they consolidated, regrouped and became confident enough to venture out and seek some sort of revenge. After the immediate outpouring of grief people probably just did not talk about the massacre. But this did not mean that they forgot and certainly some of them harboured thoughts of revenge as the Native Americans would discover to their cost, for they had given the colonists an excuse to set morality aside. Some settlers nursed a grudge and over time they would exact a terrible vengeance, but in the aftermath of the massacre the immediate concern was that the tribes would seek to finish off what they had started so action was taken to:

'Recollect the straglinge and woefull inhabitants so dismembered into stronger bodies and more secure places'

In the years that followed the massacre the settlers consolidated their position and security and encouraged new settlement. Using the argument of treachery as a justification, the colonists went on the offensive against the Native Americans:

'employing many forces abroad for the rooting out of severall places, that we may come to live in better security doubting not but in time we shall drive them from these parts'

The culture and associated mind-set of the tribes in the Powhatan Confederacy would have been very different from that of the settlers. In their value system the tactic of subterfuge and deception seems to have been an accepted battle ploy, to which connotations of shame betrayal and deceit were not attached. The Powhatan Confederacy seem to have used the tactic before in delivering retribution to other tribes. For instance earlier in the history of the colony the Virginia Company, at the time anxious to win favour with Chief Powhatan, had been complicit in such a surprise reprisal raid by providing a supportive mercenary force in return for receiving some of the booty. It seems that the Powhatan frame of reference accepted that once an inter-tribal dispute had been settled and honour satisfied then cordial relationships could be re-established without rancour. This was certainly not a view shared by Dr Potts the physician to the Virginia Company, who, in seeking revenge, apparently adopted a 'wolves in sheep's clothing' methodology similar to the one perpetrated during the massacre attack. Sometime later during the celebrations following a parley with the leaders of the Powhatan Confederacy, Dr Potts apparently administered a sleeping draught and then murdered many of his unconscious guests. Opechancanough was believed to have been present but managed to escape. Answering to the charge to the Virginia Council in 1624 Potts indignantly denied the charge but rather revealingly pointed out that a peace had not been properly concluded:

> *Besides the act is made more prodigious by the number two hundred they said to be destroyed by poison at a feast when there was neither feast nor any man poyson'd ... and fiftie said to be shott to death when there were but nine though ... we should not have doubted but the act had been justifiable.*

Reading between the lines it is clear that something nasty took place, especially when Dr Pott's dubious reputation is taken into account. (Dr Potts subsequently acquired an unsavoury reputation in part due to the testimony of one of a group of womenfolk captured and enslaved by the Native Americans during the massacre. Her husband had been indentured to Dr Potts and though he had been killed during the massacre, Dr Potts insisted that his widow serve the two year period which remained outstanding on her husband's contract. She apparently commented that it would have been preferable to return to the Indians).

In the long run the effect of the massacre on the Virginia Company was significant and devastating. After the visit of Pocahontas to England in 1616 the Virginia Company's star had been in the ascendant, now after the massacre it came crashing to earth. The shame of being so unprepared that the colony was able to be decimated by those they referred to as 'savages' would not have gone down well in the corridors of power back in England for it not only represented a catastrophic loss of face for the Virginia Company but would have been seen as a national humiliation. Though the Company continued to operate with declining authority for a further three years its fate was probably sealed from this moment. In the months after the massacre Richard and Isabel and their son George may have re-occupied their original holding at Jamestown. Richard must have fretted that after all his preliminary work at Paces-Paines he had been forced to abandon his holding He may even have doubted whether he would ever be able to return and it may have been during this time that he began to discuss his new business venture with Captain Powell. However the

Aftermath & Rebirth

security position improved and in October 1622 he petitioned the Governor Sir Francis Wyatt (in the document shown below, possibly written by Richard himself) for permission to return. This was eventually granted when word was received from England.

CCLX. RICHARD PACE. PETITION TO THE GOVERNOR AND COUNCIL IN VIRGINIA

BETWEEN OCTOBER, 1622, AND JANUARY, 1622/23

Manuscript Records Virginia Company, III, Part ii, Page 58
Document in Library of Congress, Washington, D. C.
List of Records No. 366

To the right Wor^{ll} Sr Francis Wyatt knight ec and to the rest of Counsell of Estate here

The Humble petition of Richard Pace Humbly sheweth Whereas yor petitioner heretofore hath Enioyed a Plantation one thother side of ye water, & hath bestowed great Cost & Charges vppon building ther, & Cleareing of ground but at lenght was Enforced to leaue ye same by ye sauidge Crewelly of ye Indian€. Yett now purposeing (by gods assistance) to fortifie & strengthen ye place wth a good Company of able men, hee doth desier to inhabit ther againe, & by yor leaue freely to Enioy his said plantation, promising to Doe all such thinge as by yor wo^{r}ps dyrections hee shall be Enioyned, either for ye better safe guard & defence of ye people, y^t hee shall ther put ou^r, or in w^{t}eu^r yor shall please to Comaund him

In tender Consideration Wherof may itt please yor wo^{r}ps to grant him his request, and hee shalbe bound to pray for yor health and happines both in this Worlde & in ye worlde to Come

This petition graunted, as many others also resouled vpon ther plantations according to order receaued from England

So probably by January 1622/23, some nine months after the massacre, Richard and Isabel returned to Paces-Paines. It is likely his Ancient Planter comrades who had been with him at the time of the massacre returned around the same time, with the addition of William Proctor and his wife Alice. Proctor had been in England giving evidence to a Virginia Company inquiry at the time of the massacre. He had held land at Henrico City but this had been very reluctantly abandoned by his wife in the aftermath of the attack. He fits into the pattern of the homogenous grouping around Richard for he was also an Ancient Planter who had been wrecked on Bermuda in 1609 and a firm friendship with Richard and Isabel may have dated from this time.

Richard died very soon after he returned to his holding, probably around February 1622/23. Given that by now he must have been immune to many of the common causes of death in Virginia, the likelihood is that he had a heart attack from overwork in trying to re-establish the viability of his holding. He was 42 and may even be buried in the field by the bank of the River James. Within a month or two Isabel, now aged 34, married their close friend and neighbor William Perry and almost immediately became pregnant for an infant Perry named Henry appears in the 1624 muster. Such a swift union after being widowed was not uncommon and was almost a necessity for the land had to be worked. Her new husband was a widower, for William Perry a mariner of Poplar had been married to Elizabeth Withers, a 'mayde', at St Dunstans on November 30th 1618. (It may be of significance that a Walter Rolfe was married earlier in the month in the same church). A curious fact is that when Richard and Isabel married on October 5th 1608 it is very likely that Mr Doctor Gouldman officiated at the service and he remained the incumbent vicar when ten years later William Perry married his first wife. William Perry became a significant figure in the colony and served on the Governor's Council from 1632 until his death in 1637. He must have shared Richard's concern for the indigenous population for on a return to England in 1624, he took a Tappahannah Indian boy with him and asked for funds to be used to bring the child up in the Christian faith. His son Henry inherited his property. It is very likely that Richard's kinswoman Ursula Clawson – the 'thirteenth mayde' – died at most within a year or two of her arrival in Virginia, for she does not appear in subsequent records.

After the death of her second husband at the age of 50 Isabel married a Jamestown merchant George Menefie, reputed to be the richest man in the colony. He was an attorney who traded in land and tobacco and who was also a member of the Council. Isabel died around 1645. Isabel's second son, Henry eventually married George Menife's daughter and later became a member of the Council. George Pace married Sara Maycock the daughter of Samuel Maycock who had been killed in the massacre. It is possible that she had been raised by Isabel after the death of her parents. Unusually, on her marriage to William Perry, Isabel retained ownership of land held in her own name and George later also claimed the 300 acres head right grant for the six people who had arrived on the *'Marmaduke'*. Isabel and George later made several land trades and eventually sold off that property. With George's marriage came the nearly 2000 acre plantation that Sara inherited from her father and they continued to live in Virginia. The couple had four children Richard II, John, Elizabeth and Thomas which though these are common names they can be found in the Pace family in Kingston-on-Thames. Richard II married Mary Knowles in 1661 and had eight children.

On his return to Virginia from England John Rolfe had been appointed as Secretary and Recorder General of Virginia. He married for a third time and his wife Jane Peirce had a daughter Elizabeth. John Rolfe made his will on March 10th 1621/22 just before the massacre. He may have died of natural causes for there is no mention of his murder in the subsequent records. In 1632, Thomas aged just 17, the son of John Rolfe and Pocahontas married Elizabeth Washington at St James church, Clerkenwell. The couple had a daughter Anne but Elizabeth died shortly after the birth. Thomas returned to Jamestown in 1635 to claim his inheritance and emulating his father, he left his first born daughter with his cousin Anthony Rolfe. Anne lived until 1676. In Jamestown Thomas married Jane Poythress and they had a daughter named Jane. Isabel would have been present in Jamestown when Pocahontas gave birth in 1615 and would have been there to welcome Thomas on his return twenty years later. The children of Richard Pace, William Perry Isabel Pace/Perry, John Rolfe and Pocahontas, and the Reverend Maycock all went on to become prominent citizens of Virginia.

As Richard lay on his death bed he may well have reflected on an unregretted life. When he and Isabel had originally set off, possibly as covert Puritan evangelists, they must have been prepared to accept the compromise of an Anglican form of service. It may be that they circumvented any misgivings by the simple expedient of deciding to leave the matter to the Almighty. The Bermuda Crest bears the inscription *'Whither the Fates Lead Us'.* Richard and Isabel may well have adopted a philosophy of whither God leads us. They may well have been content as devout Christians, to surrender themselves with implicit faith and trust to God's goodness and mercy. At the end of his life Richard would have concluded that God had indeed answered his prayers and protected his family and walked with them through all the trials and tribulations they had faced. He may have believed that he had perceived what he interpreted as many miracles or near miracles. The saving of all aboard the *'Sea Venture'* when she foundered off Bermuda was indeed miraculous. Bermuda was known as the 'Island of the Devils' but the uninhabited land had provided sustenance and the wherewithal to construct two ocean going ships in which to complete the voyage to Virginia – an extraordinary achievement. Arriving a year overdue they had avoided the 'starving time' during which the family may well have perished. Another near miracle was perhaps meeting the De La Warr 4th Supply voyage off Mulberry Island on the very day that Sir Thomas Gates had abandoned Jamestown. The 4th Supply must have been sent as a replacement for the *'Sea Venture'* that had been presumed lost. De La Warr must have been astounded to find Sir Thomas Gates alive but had he found no one at Jamestown he would surely not have stayed. After the death of John Rolfe's wife and child on Bermuda, his subsequent marriage to Pocahontas ended the 1st Powhatan War and this may have been seen as a divine intervention. The experimental development of a favourable strain of tobacco which became the staple crop that eventually allowed the colony to prosper may similarly have been attributed to a divine force. Richard and Isabel would have thanked God for the turnaround in their own life chances following the grant of 100 acres each for their personal Adventure and for their safe double crossing of the Atlantic to recruit a workforce for their plantation for which they gained an additional 300 acres. The warning given by Chanco on the eve of the massacre attack followed by the successful delivery of a warning to Jamestown aided by a fortuitous tidal flow and reasonably clement weather may also have been seen as a near miracle. Richard and the small group of friends around him probably all shared the common goal of seeking God's grace through rescue, redemption and liberation. Their fervent hope would have been for everlasting life in the House of the Lord. In the light of this history and the survival of his family Richard and his hardcore of devout Ancient Planters may even have concluded that indeed, they were among those who may have been chosen by God.

William Shakespeare was a near contemporary of Richard Pace. Celebrating the 400th anniversary of his death in April 1616, the Vicar of the Church of the Holy Trinity Stratford on Avon the bard had attended as a boy, described how he would have been given a good deal of religious instruction at home as well as sitting through innumerable sermons in the church. He judged that Shakespeare would have been very familiar with the New Testament and particularly the Book of Job and the Psalms from the Old Testament. Richard must also have been very familiar with these texts. An orientation towards the teachings of the New Testament might suggest Richard would have rated love

Aftermath & Rebirth

and forgiveness over justice. If so, then after the massacre he may well have advocated merciful arrangements with the Powhatan Confederacy suggesting reconciliation rather than retribution. At the end of his life he may well have recalled A Psalm of David which he may have seen as a remarkable resumé of his experience of life. Richard and Isabel had survived tyranny, semi slavery and hunger in Virginia, but together with their young son George they had survived. He would have recognised that he and his family had walked through the valley of the shadow of death but had feared no evil and that God had indeed protected and comforted them. He would have surely reflected that God's goodness and mercy had followed him all the days of his life. Richard probably had an unregretted death. However strong his family ties, for a devout Christian such as he was, he would have believed that his true home belonged in another realm and until he reached it through death he would have considered himself an exile from the family of God. His fervent hope would have been to dwell in the House of the Lord forever.

Richard Pace must have been a restless individual. He could have chosen to work as a jobbing Carpenter in London He may well have lived longer and fathered a larger family if he had stayed at home. Instead he and Isabel chose to adventure to the New World. The more rewarding purposes of human existence might be thought to include living a good and virtuous life, in seeking spiritual and intellectual enlightenment, helping and teaching others and the opportunity to be creative. Perhaps the greatest achievement of humankind is to love and be loved by others. The contemporary reports of Richard's warning to Jamestown are matter of fact and do not eulogize him. The ostensible reason for this is that Richard had simply rowed across the James River with news of the impending attack. But the actual root reason why the Virginia colony survived was almost certainly because gentle Richard had treated an indigenous American Indian servant as a son, who in return had loved him as a father. A massacre at Jamestown Fort was foiled by Christian love. Richard Pace would probably be content to have that as his epitaph.

APPENDIX

The Author

David Edmund Pace born 1940, is a retired lecturer in management studies who spent his early career at sea. He attended the University of East Anglia and also studied at the Department of Occupational Psychology, Birkbeck College, London University. He has spent many years researching the history of the Pace family.

Other books by the author:

Direct Participation in Action: the new bureaucracy (with John Hunter)
Saxon House

A Quaker Family from the Severn Vale: Volume I The Quickening. 1500-1750

Volume II Timekeeping 1750-2000

Research undertaken for this publication is available

The Clerkenwell Pilgrims (2014 &2015)

These publications are available on CD from the Pace Society: info@pacesociety.org

See also: 'The Pace family of Quaker clockmakers' – Antiquarian Horology Number One Volume Thirty-Four, March 2013

Authors Note

For over 20 years the Jamestown Rediscovery project has delved into the early history of the site of the first permanent settlement on American soil. By providing a constant stream of information and updates to all those fascinated by their excavations the archaeologists have managed to transform what it is inevitably a slow, meticulous, and painstaking process into a flow of fresh and exciting theatre. So for instance when a halberd with a bent axe head engraved with griffins – the emblem of the De La Warr family – was recently discovered in the well at James Fort the interpretation was riveting. Excavation, restoration and identification, historical records, the find location and creative thinking all contributed to the conclusion that the weapon had probably been deliberately deformed into a long handled tool to recover an item that had fallen into the well. It is possible to personally identify with this predicament and the story brings immediacy to the situation.

In writing this book similarly I have been much concerned to avoid producing an historical tome and it is mainly for that reason I have not included any notes or academic references, but it is based on considerable research. This is a truly epic story which I have tried to present as a narrative. If I have failed to do it justice then the fault is mine and in no way should it detract from the homage due to the extraordinary people who first founded America. This book is not intended to be the story of the colonisation of Virginia but the story of one man's experiences during the early period of colonisation. It is a micro rather than a macro account and an attempt to provide an accurate, organised description of the first American settlement.

Several categories of information have been used to build the narrative. The background setting is the zeitgeist of the period – the deep religiosity conditions everything – the divisions in the Anglican church that developed in the decades following the Reformation, the compulsory church attendance, the desperation to be buried in consecrated ground and the sheer funk about the possibility of purgatory. This was the contemporary mind-set that is difficult to comprehend in today's world but without a thorough grasp of this any meaningful understanding of Richard and Isabel's actions and commitment is impossible. I have drawn upon Professor Tomb's recent comprehensive summary to describe this which gives the essential backdrop to the story.

The next source has been the hard information drawn from official records such as Parish Registers in England and the records of the Virginia

Company. These have allowed the chronology of events to be established, the quantification of such aspects as mortality rates to be assessed, and sometimes to give insight into personal feelings. For instance, Lord De La Warr is clearly very anxious to convince his peers that his early departure from the colony in early 1611 was due to genuine ill health and near death and similarly John Rolfe is worried that leaving his son behind when he returned alone to Virginia might be construed by his equals as abandonment. There is also a report of Chief Powhatan's reaction on hearing of the death of his daughter Pocahontas and his grandson's sickness.

Further sources are the relatively few personal accounts of what happened. The most significant of these used here is the Ancient Planters' Declaration on which Chapter 3 is based. This is a contemporary account of the dire circumstances in which the settlers were forced to survive, the lack of supplies, the despotic regime and the harsh punishments inflicted. Similarly William Strachey's account of the foundering of the *'Sea Venture'* on Bermuda taken from a personal letter is both graphic and beautifully written. The only surviving document possibly written by Richard Pace is his submission to return to his holding some 6 months after the massacre.

Writing an account 400 years after the event means it is necessary to interpret what happened. In order to link these facts into a narrative I have had to rely on the 'likeliest' choice. The records may signpost but any uncertainty must be resolved by informed judgement, not guesswork, having first asked some pertinent questions. So for instance, in Chapter 1 the question arose of why Richard was baptised in Kingston-on-Thames and his brother was baptised 4 years later at Farnham 30 miles away. Only when data was accessed from the local Surrey Family History Society – data that had not shown up on the on-line Parish Register excerpt – which showed that Richard of Jamestown's father Richard had died in 1587, did the likely course of events become clearer. At that point the plight of his wife Anne Browne, a young destitute widow with two young sons now becomes understandable and the judgement was made that she probably sought sanctuary with her sister Johan Browne married to the carpenter John Clawson in East London. It seems probable that the younger son Thomas was adopted by his uncle John at Kingston who had no male heir, and eventually inherited the family holding there. Again a small detail from the original record at the Guildhall Library revealed that John Clawson was the 'supposed' father of his son Robert (the probable father of Ursula Clawson who sailed to Virginia in 1621) – again a fact that does not show up in on line

Author's Note

records. This again leads to question why – Was Johan a fallen woman? Had she endured a rape attack that had left her pregnant? Far more likely is that she was a pregnant wife whose husband had suddenly died – and Kingston records suggest that this might be so – though one has to live with a degree of uncertainty.

The most controversial conclusion in the account is the claim that Richard and Isabel were aboard the *'Sea Venture'* when she was wrecked. A judgment has been made that the couple resolved to emigrate to Virginia at the time of their marriage in October 1608. The 3rd Supply fleet – the largest civilian convoy ever assembled – sailed the following year in May 1609. Richard almost certainly knew John Rolfe as they probably worshipped at St James Clerkenwell where John Clawson the carpenter was buried in 1598 and where Thomas Rolfe would eventually marry. With all the publicity and excitement surrounding the departure it would be the likeliest time to depart with their friend John Rolfe and his pregnant wife. An elephant in the room throughout the 3 year study has been to answer the question as to why Richard named his son George. This is seemingly out of sequence and seems to have no obvious family connection. The judgement has been made is the likeliest reason is that he was named after Admiral Sir George Somers whose brilliant seamanship was primarily regarded as being responsible for saving all 150 on board after the ship sank. Only if the ship's manifest ever comes to light will it be possible to learn the identity of the other 100 unnamed people on board. A further clue is that shortly after news of their miraculous survival was received in London, and more than a year after being reported missing presumed lost, George Davison a parishioner at St James, Clerkenwell named his newly born twins Richard and Isabel – perhaps in honour of his friends who had returned from the dead!

The final source for the text has been the imagination – trying to put oneself in the position of Sir George Somers as he fought to save his ship, supported by the evidence of Strachey's account – or trying to put oneself in to shoes of Richard Pace as he launched his boat and rowed across the James River to warn Jamestown of impending attack – again with the account supported by precise calculation as to tidal flow and moon phase. Possible weather conditions have been judged by modern day forecasts for the appropriate month. Just as the archaeological excavations and the items recovered over the past twenty years at Jamestown Rediscovery have enabled the everyday life of the early settlers to be reconstructed, imagination can transport one back to create an almost present reality and allow almost a lived experience and a direct connection to the past. The sense of immediacy and attachment to these momentous events can also be traced back in England.

This book is being written from a vantage point high up on the White Cliffs with the backdrop of the Roman lighthouse and the Norman castle overlooking the Straits of Dover. Had I been here in May 1609 I would have been able to see the nine ships of De La Warr's 3rd Supply voyage sailing west led by the *'Sea Venture'* under the command of Admiral Sir George Somers carrying John Rolfe and his pregnant wife Elizabeth and very probably Richard and Isabel Pace to Virginia. In 1616 I could have spied the *'Treasurer'* bringing John Rolfe and his wife Pocahontas together with their infant son and accompanied by Sir Thomas Dale sailing past. In August I may have seen the widowed John Rolfe returning, for 50 miles away over my right shoulder Pocahontas still lies under the chancel at old St George's church at aptly named Gravesend. Richard and Isabel sailed past this point in 1621 on their way to recruit a workforce for their newly acquired plantation and returned a few weeks later in the 100 ton *'Marmaduke'* accompanied by his teenage kinswoman Ursula Clawson and the first batch of maids sent out as potential brides for lonely settlers. I also know what befell these hopeful settlers in the intervals between these sightings and also the tragic fate of the great majority of them. Almost all of those who sailed in the great fleet of 1609 who were not on the *'Sea Venture',* would be lost at sea, murdered or die of starvation having experienced the horror of cannibalism. Pocahontas would never see her homeland again, John Rolfe would never see his son after leaving the sickly boy at Plymouth. Nearly all the innocent maids shipped out were doomed to an early death. Richard and Isabel were the great survivors.

STEPNEY REDISCOVERED

Visiting a location can add a dimension to a lived reality. Metaphorically it is possible to link the two places where Richard and Isabel predominantly dwelt. Let me take you for a stroll through Stepney in the East End of London. Make sure you have purchased a travel card with your railway ticket to St Pancras International Station. There take the Hammersmith and City Line tube to Stepney Green, (or the District Line from elsewhere in London). As you come out of Stepney Green tube turn right cross the busy road walk for about 300 yards before turning left into Stepney Green Road. Follow this south – in to the sun if you are fortunate – and as you round the bend you will see St Dunstan's Church before you, standing in its railed green. You will be surprised

Author's Note

at its size for in Richard's time it was the only church in the vicinity and it had to serve an extensive area. There is the Vicar the Rev. Doctor Gouldman bidding farewell to Richard and Isabel and giving a final blessing after their marriage on 5th October 1608. Ten years later the same priest would officiate at the first marriage of William Perry who would become Isabel's second husband in 1622/23. The road line would likely to have been identical in 1608 so come out of the church gate and turn left retracing their steps south towards the river about 10 minutes away. Traverse Commercial Road and then cross the busy Highway and turn right for 300 yards. I am familiar with this area for as a 16 year old I was at nautical college close by in the Commercial Road and went to sea for the first time in the college training schooner sailing from Wapping Basin. At the blue signpost for Shadwell turn left, pass the Shadwell Basin Dock on your right and directly in front you will see the ancient 1520 pub the *'Prospect of Whitby'* in Wapping Wall.

The pub backs directly onto the River Thames which in Richard's day would have been lined with shipbuilders. It is unlikely that Richard and Isabel stopped here after their marriage for a celebration with friends for then it was known as the *Devil's Tavern'* and had a dubious reputation. Walk a few yards along

Wapping Wall and a real surprise shock for there is a cul-de-sac to the right called Pear Tree Close. Richard's cousin Robert Clawson's eldest son lived in Pare Tree Alley in 1630 – surely too much of a coincidence. This may well have been where the family house had been built for in 1596 Elizabeth I had given a dispensation that allowed shipyard workers to build on the river bank. Before you leave cut through the 10 yards to the river. You will be surprised how wide the Thames is here and you will be able to appreciate how convenient a highway it would have been in those days. Perhaps those ancient stumps sticking out of the mud was once a jetty used by Richard. What becomes very apparent is the obvious fact that Richard, who at the time of his marriage was known to be living at Wapping Wall, would almost certainly have been working on the shipyards on the river bank.

Perhaps the vessels returning from Virginia such as the *'Godspeed'* came in for a refit and it was from the sailors that Richard first learned the opportunities opening up in Virginia. A few yards further at the end of the road there is Wapping Overground station for which your travel card is valid to whisk you back to the City with a change at Whitechapel. Allow a maximum of 3 hours although you may choose to tarry a while for lunch at the *'Prospect of Whitby'*.

Author's Note

Training/Schooner 'Wendorian'
Wapping Basin, March 1957.

It was from Wapping that both the author and Richard Pace set off to sea.
(The dock no longer exists)

Note: By chance in my subsequent career I continued to retrace Richard and Isabel's path, for the shipping company that employed me traded on the old triangular route from England to West Africa, across the Atlantic to the Eastern seaboard of the USA and then via the northerly route home. When sailing on a voyage from West Africa to the USA in 1958 I was severely bitten while trying to recapture a cageload of escaped monkeys and developed a high fever. Fearing for my life the captain altered course to hospitalize me in Bermuda – so curiously I shared with Richard a near death experience in the vicinity of that Island. (In the event the captain changed his mind and the ship carried on to New York). A month later our departure point was Norfolk, Virginia – although at the time I had no inkling of the historic events that had taken place just around the corner at Jamestown.

David Edmund Pace
Dover. July 2016

www.ingramcontent.com/pod-product-compliance
Lightning Source LLC
LaVergne TN
LVHW021119080426
835510LV00012B/1755